Grasshoppers and Crickets
of Surrey

Grasshoppers and Crickets
of Surrey

DAVID W. BALDOCK

SURREY WILDLIFE TRUST

Cover illustration: Rufous Grasshopper (nymph), by David Element

ISBN 0 9526065 4 2

British Library Cataloguing-in-Publication Data.
A catalogue record for this book is available
from the British Library.

© David W. Baldock 1999
Surrey Wildlife Atlas Project

All rights reserved.
No part of this publication may be reproduced or transmitted in any
form or by any means, electronic or mechanical, including photocopy,
recording, or any information storage and retrieval system, without
permission in writing from the publisher.

First published 1999
by Surrey Wildlife Trust
School Lane, Pirbright, Woking, Surrey GU24 0JN.

Printed and bound in Great Britain by
Biddles Ltd, Guildford and King's Lynn

FOREWORD

The human residents of Surrey are lucky to live in one of the most attractive counties of southern Britain, including such beauty spots as Box and Leith Hills, the Hog's Back, the Devil's Punch Bowl and Frensham Ponds. The county straddles the North Downs and so benefits from a generous portion of the London Basin, overlying Tertiary geology, to the north of the chalk downland, and an equally large area of the Weald, overlying Cretaceous geology, to the south. The chalky soils that lie across the middle of the county support a rich lime-tolerant flora and fauna, in contrast to the acidic soils to the north and south, where there are areas of both dry and wet heathland. Water-meadows are provided by the rivers Wey and Mole, and there is an abundance of natural woodland.

This variety of habitats has resulted in an unusually rich orthopteroid fauna, including no fewer than 30 out of a possible total (including earwigs) of 35 species. David Baldock, with the help of Roger Hawkins for the eastern part of the county, has spent some 30 years in an astonishingly thorough study of their distribution, and the impressive results can be seen in the maps illustrating this book, together with the most interesting and readable text that accompanies them. One of the remarkable facts to emerge from this survey is that some species have increased their range very rapidly during the past 10–20 years. When I first became interested in Orthoptera in the forties I had to travel to Dorset to be sure of finding the Long-winged Cone-head; now this species has spread to a large area of southern England and occurs quite commonly in most of Surrey, where it was unknown until 1990. Roesel's Bush-cricket was limited in Surrey to one or two localities in the east of the county until the seventies, when it began to spread westwards; it now occurs in localities scattered throughout the county. As the author explains, tracing the spread of these bush-crickets has been greatly helped by the use of a bat-detector to find them by the calling songs of the males.

David Baldock's book will be an essential guide for all those interested in the Surrey orthopteroids. It is sure to stimulate more local interest in these insects and will no doubt result in the addition of further records – perhaps even the addition of a new species to the Surrey list.

DAVID R. RAGGE

To my grandson
BEN

CONTENTS

Preface 1
Acknowledgements 3
List of recorders.. 5
Introduction
    Orthoptera – structure and life history 6
    Classification 9
    The present status of Orthoptera in Surrey 10
    Surrey – the survey area 15
    Climate, geology and distribution 17
    The fossil record 20
    History of recording in Surrey 23
    Identification 28
    Songs and bat detectors 30
    Explanation of species accounts and distribution maps 37
    Further reading 39
    Future recording 40

Species accounts
    Bush-crickets 43
    Crickets 63
    Mole-crickets 70
    Groundhoppers 74
    Grasshoppers 77
    Cockroaches 90
    Earwigs 94

Appendices
    1. Gazetteer of sites 99
    2. Organisations 103
    3. References 104
    4. Glossary 108
    5. Plant names 110

Index
    Latin 112
    English 112

"Men that undertake only one district are much more likely to advance natural knowledge than those that grasp at more than they can possibly be acquainted with: every kingdom, every province, should have its own monographer."

GILBERT WHITE

PREFACE

Surrey is still one of the finest counties in Britain for most groups of insects, and the grasshoppers and crickets are no exception; they are almost as well-represented in Surrey as in the prime counties of Dorset and Hampshire. This is in spite of of the fact that Surrey is entirely land-locked, unlike most of the best entomological counties, and in spite of its close proximity to London; indeed most of the north-east of the vice-county is within the former Greater London. The main reasons for this richness of species are the variety of the underlying geology, which in turn affects the habitat diversity, and the large amount of natural and semi-natural landscape which still exists throughout the county. Another reason is that Surrey, situated as it is in the south-east corner of England, enjoys a warmer and drier climate than most other counties. Most of the grasshoppers and crickets are warmth-loving insects, preferring undisturbed, dry and mainly open habitat, and in Surrey there is still plenty of such habitat.

Although the title of this book is *Grasshoppers and Crickets of Surrey*, we have included in the survey the native cockroaches and earwigs, as these have traditionally been treated together by earlier authors, and were until recently included in the Order Orthoptera. They now have their own Orders, the cockroaches Dictyoptera and the earwigs Dermaptera. All the insects dealt with in the book are therefore referred to as orthopteroids. Its more correct title should strictly be, either 'Orthopteroids of Surrey', or 'Grasshoppers, Bush-crickets, Crickets, Cockroaches and Earwigs of Surrey', but neither of these titles commended themselves to the publishers!

This is the fifth in the series of county atlases covering various insect groups in Surrey. Although I hope it will appeal to the specialist entomologist, because of the distribution maps and the many previously unpublished observations, I also hope that it may encourage others with little knowledge of insects to become interested in this fascinating group. For this reason I have kept Latin names and technical terms to the minimum; fortunately most species have had well-established English (or vernacular) names for many years, and in some cases for centuries.

Grasshoppers and their relatives make an ideal group of insects to study and survey for the following reasons:

– They form a small group of only about 35 species countrywide.

– They are mostly large and therefore easy to find, two of them being amongst the largest of all British insects.

– Between them they have a very long season, from early spring right through to late autumn; some can even be found, as nymphs, in winter.

- Most are easy to identify in the field without even a field-lens, and there is therefore no need to keep specimens.
- Most of the commoner species are ubiquitous and often abundant; there is no tetrad in Surrey, even in Central London, without at least one species.
- There are good, modern, well-illustrated and readily-available books on the group.
- There is a very active National Recording Scheme, which produces a newsletter annually, and which has regular features in *British Wildlife* magazine.
- Their songs make recording for a survey much quicker than with other insects, since there is no need to catch or even see most of them.
- They are ideal subjects for photography as many species remain motionless for long periods.

This survey has shown that the orthopteroids are thriving in Surrey at the end of the millennium. Five new species have been discovered in the last 50 years or so, three of them newcomers, and two other species appear to have increased their range considerably in the last decade or two; most of the others are as common as they ever were. Only one species, the Field Cricket, has been lost, although it is now being reintroduced, and another, the Stripe-winged Grasshopper, is possibly losing ground due to habitat changes.

One object of this survey is to show, as precisely as possible, the distribution of these insects at a certain time, i.e. at the end of the millennium, and hopefully this atlas will be used as a benchmark for all future study of this group. As Edmund Jarzembowski explains in his chapter on fossils, the orthopteroids were flourishing in Surrey with the dinosaurs about 125,000,000 years ago and they look likely to be with us for a long time yet.

ACKNOWLEDGEMENTS

No county survey can be carried out without the help of many people and I have been fortunate in the help which I have received. Numerous people have sent in records and I am indebted to all of them; a full list of recorders is given later in the book.

But this book would never have seen the light of day without the constant help and effort of Roger Hawkins; his contribution to the survey as well as to the book has been almost as much as mine. As soon as he arrived in Surrey in 1976, he offered to record in the south-east, which at that time was greatly under-recorded. Over the next 20 years he covered the whole of the eastern half of the county, as well as making many forays into the west, especially in the north-west corner which was also under-recorded. He was probably the first person in Britain to use a bat-detector for recording species distribution, and this method revolutionised our speed of recording in the latter half of the survey. Not only was he recording, but he was also making meticulous notes of his observations of the songs, dates and habits of orthopteroids in the field, in particular of the bush-crickets, and many of these observations are included in this book. In addition to all this he has encouraged me throughout to write and publish this book; he has made countless constructive suggestions as well as making numerous corrections and alterations to the text, and finally did the proof-reading. My gratitude to him is enormous.

I would also like to thank the following people for their help: Graham Collins, for preparing the final version of the distribution maps and for his helpful suggestions on early drafts of the text, as well as for his constant advice on computer use; his father Geoffrey Collins, for giving me valuable advice, particularly on pre-1970 records; David Ragge, for reading an early version of the text and for giving me constructive suggestions, as well as kindly allowing me to reproduce the song diagrams and drawings from his book and for writing the foreword to this book; Chris Haes, for his constant enthusiasm and encouragement from the very start of the project, and also for refereeing the species accounts; and Mike Thurner, for preparing the original distribution maps. These maps have been produced by the DMap Program (in its Windows version) written by Dr. Alan Morton of Imperial College at Silwood Park.

Surrey is as rich as any county in its fossil orthopteroids, which are found mainly in the claypits of the Surrey weald, e.g. Smokejacks, south of Walliswood, and Auclaye and Clockhouse, south of Capel. Because of this I asked Edmund Jarzembowski to write a short chapter on the fossil record, and I wish to thank him for this and the extraordinary photographs of 125,000,000 year old insects.

I also wish to thank John Widgery, Graham Collins, Mike Thurner, Bryan Pinchen, John Patmore, Mike Edwards, Jim Porter, Roger Key, Chris Haes and, in particular, David Element, for donating the superb slides used to produce the colour-plates in this book.

Finally I want to express my appreciation to Paul Wickham and Clare Windsor of Surrey Wildlife Trust, which is the publisher of this series of books: Paul Wickham for his enthusiasm in keeping the publication of this series on schedule and Clare Windsor for preparing this book for printing and publication in such style.

I have dedicated this book to my grandson Ben, who, at the age of only one, accompanied me with great enthusiasm on grasshopper hunts, at the age of five was advising me on the use of my computer, and at the age of seven was helping me to analyse the frequencies of bush-cricket songs.

DAVID W. BALDOCK

LIST OF RECORDERS

The following people have contributed records, from one to many hundred, and where records are given in the text the recorders are normally indicated by their initials. Recorders are listed in alphabetical order of their initials to enable easy identification.

AC	A.Channer	JBS	J.B.Steer
AJH	A.J.Halstead	JMcL	J.McLauchlin
AJP	A.J.Pontin	JRD	J.R.Dobson
ARGM	A.R.G.Mundell	JP	J.Porter
CM	C.Martin	JPW	J.P.Widgery
CRH	C.R.Hall	JSD	J.S.Denton
CWP	C.W.Plant	KNAA	K.N.A.Alexander
DAC	D.A.Coleman	ME	M.Edwards
DE	D.Element	MJS	M.J.Skelton
DRR	D.R.Ragge	MSP	M.S.Parsons
DT	D.Tagg	MT	M.Thurner
DWB	D.W.Baldock	RAJ	R.A.Jones
ECMH	E.C.M.Haes	RBH	R.B.Hastings
GAC	G.A.Collins	RDH	R.D.Hawkins
GBC	G.B.Collins	RH	R.Holder
GLAC	G.L.A.Craw	RKAM	R.K.A.Morris
GNG	G.N.Gardiner	RMF	R.M.Fry
HCE	H.C.Eve	RMcG	R.McGibbon
HGI	H.G.Inns	RWJU	R.W.J.Uffen
IS	I.Saunders	VKB	V.K.Brown
ISM	I.S.Menzies	WRD	W.R.Dolling
JAM	J.A.Marshall		

ORTHOPTERA—STRUCTURE AND LIFE HISTORY

Orthoptera (from the Greek words meaning 'straight wings') form a large group of ancient, primitive insects containing about 20,000 species worldwide; they are mainly warmth-loving insects and therefore occur most commonly in the tropics and only about 600 species are found in Europe, while in Britain, which is on the edge of their range, only 35 have been recorded, even after including those insects that have been split off from the old order Orthoptera into separate Orders. The group includes such different insects as grasshoppers, crickets, bush-crickets, mole-crickets, cockroaches, earwigs, praying-mantises and stick-insects. The few characteristics which they all have in common include their biting mouthparts and their forewings which are thicker than their very delicate and foldable hindwings; these forewings protect the vulnerable hindwings when at rest.

Being primitive, they all have a fairly basic anatomy and lack the complete metamorphosis of most insects such as butterflies, bees, flies, beetles, etc. They nearly all emerge from the egg as a small worm-like (vermiform) larva which almost immediately sheds its cuticle and becomes a small version of the adult, known as a nymph. Earwigs and stick-insects do not even have the larval stage but emerge from the egg as nymphs. The nymph passes through various stages of growth, called instars, moulting its skin between each, until it eventually becomes a sexually mature adult. The adult female then mates and lays eggs and the life-cycle starts again. In most British species the life-cycle is completed in one year, but the native cockroaches and some bush-crickets take two years to develop.

Each of the different Orders within the Orthoptera group, i.e. grasshoppers and bush-crickets, earwigs and cockroaches, have varied details of structure,

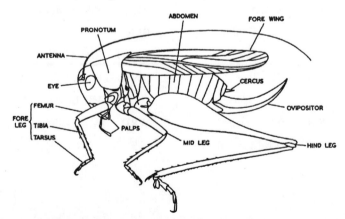

The general structure of a female bush-cricket

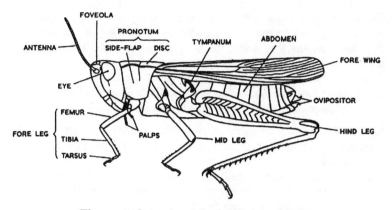

The general structure of a female grasshopper

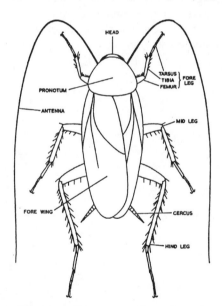

The general structure of a cockroach

as well as differing life habits. It is not appropriate in this book to go into all these details and the reader should consult one of the reference books mentioned later. However the general structure of a typical bush-cricket, grasshopper and cockroach is shown here so that the reader can recognise the various parts of the insects. These drawings are reproduced with the consent of David R. Ragge from his book *Grasshoppers, Crickets and Cockroaches of the British Isles*.

Grasshoppers and bush-crickets are best known for their ability to sing; the different songs and methods of song production are dealt with in a later chapter. Because they sing they also need to be able to hear the songs of other grasshoppers; the females are attracted to the males by hearing their songs. This means that they must have hearing organs, known as tympana. The grasshoppers have these situated on their abdomen as shown in the drawing and the bush-crickets have them in the tibia of their forelegs.

Some of the details of the habits of the different species are dealt with in the main part of the book, under the species accounts, but there are still many details which have not yet been discovered. For instance, little is known about what bush-crickets eat in the wild; they are mostly vegetarian but some species also eat other insects and in captivity some are notably cannibalistic. Many species of grasshoppers and bush-crickets vary in colour and wing-length and some of these different forms are shown in the illustrations. But it is still not certain whether these forms are produced genetically or as a result of environmental factors. Amateur orthopterists can still reveal unknown facts by careful observation and study in the field.

CLASSIFICATION

Insects, as well as all other living organisms, are classified by taxonomists into groups, in order to try to show the relationships and differences between the various Orders, Families, Genera and Species into which they are split. The Insect Class is divided into Superorders; one of these Superorders is Orthopteroidea (Orthopteroids) or Orthoptera in a broad sense. This Superorder is now divided into three Orders: Dermaptera (earwigs), Dictyoptera (cockroaches and mantises) and Orthoptera in the strict sense (grasshoppers and crickets).

This Order Orthoptera is sub-divided into two Suborders: Ensifera (bush-crickets, crickets and mole-crickets, all having long antennae) and Caelifera (groundhoppers and grasshoppers, all having short antennae). Suborder Ensifera is further sub-divided into two Superfamilies: Tettigonioidea (bush-crickets) and Grylloidea (crickets and mole-crickets). The Superfamilies are then divided into Families which are finally divided into Species.

The classification of the present-day and fossil, native Orthopteroids of Surrey is shown below; some alien species belong to other Families.

      Superorder ORTHOPTEROIDEA (Orthopteroids)
        Order ORTHOPTERA
          Suborder ENSIFERA
            Superfamily TETTIGONIOIDEA
              Family Prophalangopsidae **(extinct Bush-crickets)**
              Family Elcanidae **(extinct Bush-crickets)**
              Family Tettigoniidae **(Bush-crickets)**
            Superfamily GRYLLOIDEA
              Family Gryllidae **(Crickets)**
              Family Gryllotalpidae **(Mole-crickets)**
          Suborder CAELIFERA
            Superfamily ACRIDOIDEA
              Family Tetrigidae **(Groundhoppers)**
              Family Locustopsidae **(extinct Grasshoppers)**
              Family Acrididae **(Grasshoppers)**
        Order DICTYOPTERA
          Suborder BLATTODEA
            Family Blattellidae **(Cockroaches)**
        Order DERMAPTERA
            Family Labiidae **(includes Lesser Earwig)**
            Family Forficulidae **(other Earwigs)**

THE PRESENT STATUS OF ORTHOPTERA IN SURREY

There are 35 species of native orthopteroid insects (including the cockroaches and earwigs) which have been recorded in mainland Britain, and no less than 30 of these have been recorded in Surrey. As a result of this survey it has been possible to make a proper assessment of the present status of these 30 species in Surrey and this is given below:

Oak Bush-cricket
Meconema thalassinum
Widespread and abundant.

Great Green Bush-cricket
Tettigonia viridissima
Only two colonies known for certain.

Dark Bush-cricket
Pholidoptera griseoaptera
Widespread and abundant but absent from some areas.

Bog Bush-cricket
Metrioptera brachyptera
Restricted mainly to damp heaths but there abundant.

Roesel's Bush-cricket
Metrioptera roeselii
Has recently become widespread and quite common, spreading from the north-east.

Long-winged Cone-head
Conocephalus discolor
First recorded 1990, but now widespread and abundant.

Short-winged Cone-head
Conocephalus dorsalis
Locally abundant in the west and spreading.

Speckled Bush-cricket
Leptophyes punctatissima
Widespread and very common.

House Cricket
Acheta domesticus
Widespread in buildings, but scarce and decreasing rapidly; rare out of doors.

Field Cricket
Gryllus campestris
Always rare and very localised, and presumed extinct since 1964. Recently reintroduced.

Wood Cricket
Nemobius sylvestris
Only one thriving colony which was possibly introduced.

Mole-cricket
Gryllotalpa gryllotalpa
Always local but possibly still surviving at very low density.

Slender Groundhopper
Tetrix subulata
Widespread but local.

Common Groundhopper
Tetrix undulata
Widespread and reasonably common.

Large Marsh Grasshopper *Stethophyma grossum*	One probably native and one introduced colony.
Stripe-winged Grasshopper *Stenobothrus lineatus*	Widespread but local, on both chalk and sand.
Woodland Grasshopper *Omocestus rufipes*	Confined to the south and south-west, but there locally common.
Common Green Grasshopper *Omocestus viridulus*	Widespread but local.
Field Grasshopper *Chorthippus brunneus*	Widespread and abundant everywhere; the commonest grasshopper.
Meadow Grasshopper *Chorthippus parallelus*	Widespread and abundant except in London.
Lesser Marsh Grasshopper *Chorthippus albomarginatus*	Widespread but very local and in places common; spreading.
Rufous Grasshopper *Gomphocerippus rufus*	Restricted to the chalk, where it may be abundant.
Mottled Grasshopper *Myrmeleotettix maculatus*	Widespread and common but almost restricted to heathland.
Dusky Cockroach *Ectobius lapponicus*	Widespread and common in the west, absent from the east.
Tawny Cockroach *Ectobius pallidus*	Very local and uncommon, mainly on the chalk, but also on sand.
Lesser Cockroach *Ectobius panzeri*	Only known for certain from one locality.
Lesser Earwig *Labia minor*	Widespread and possibly common in dungheaps.
Short-winged Earwig *Apterygida media*	Very local in the extreme east.
Common Earwig *Forficula auricularia*	Widespread and abundant.
Lesne's Earwig *Forficula lesnei*	Widespread but local.

The remaining five species which have not yet been recorded in Surrey are:-

Wart-biter
Decticus verrucivorus
Only occurs near the coast in E. Kent (as a re-introduction), E. Sussex and Dorset, and inland in Wiltshire. Very few colonies known to survive and it is being reintroduced to some of its historic sites.

Grey Bush-cricket
Platycleis albopunctata
Only occurs within a few yards of the southern coasts and therefore most unlikely to occur in Surrey, although it has recently been reported as spreading inland in Dorset and S. Hampshire.

Scaly Cricket
Pseudomogoplistes vicentae
(previously misidentified as *P. squamiger*)
Only known from three sites on coastal shingle in southern England.

Cepero's Groundhopper
Tetrix ceperoi
Almost entirely confined to the south coast, with very few inland sites. It is extremely difficult to separate from the Slender Groundhopper and it may be overlooked inland in old sandpits.

Heath Grasshopper
Chorthippus vagans
Restricted to a small area of heathland in the New Forest and east Dorset. It was only first identified in this country in 1933 and has very recently been reported as spreading, due possibly to hotter summers. It is conceivable that it may spread to west Surrey, where there is plenty of suitable habitat.

MIGRANT SPECIES

There are only two migrant species which reach Britain occasionally, both of them locusts. The Desert Locust (*Schistocerca gregaria*) comes from northwest Africa and on its very rare occurrences in the British Isles it is normally recorded from the west. One adult was found dead at Selsdon in late October 1973 by Mr. Frith (det. GBC) and, although it was mature, it was considered by the Centre for Overseas Pest Research to have been an accidental introduction or an escape from captive stock, since there were no other migrant insects recorded at that time and it was widely bred for school research. The Migratory Locust (*Locusta migratoria*) reaches Britain rather more often from southern and eastern Europe. In 1947 there was a small invasion when 32 specimens were reported from southern England, but the only Surrey record was of one adult female found at Leatherhead on 22nd September 1947 by J.E.S.Dallas. This specimen, which was retained by the Anti-Locust Research Centre, was of the gregarious phase of the subspecies *gallica* and thus probably came from the south-western district of France, around the Gironde. There is a third grasshopper which looks like a locust, the Egyptian Grasshopper (*Anacridium aegyptium*) from the Mediterranean region, but this is not known as a migrant, only as an accidental introduction. One was found at Guildford in March 1959 on imported mimosa (specimen in R.H.S. Wisley Collection) and another at a nursery garden in Carthouse Lane, Chobham on 17 December 1998 (specimen with JSD).

ALIEN SPECIES

Most of those species of alien and introduced orthopteroids (excluding the Stick-insects) which are currently breeding or have recently bred in Britain, have been recorded at some time in Surrey. These are as follows:

Greenhouse Camel-cricket *Tachycines asynamorus*	Colony in greenhouses at Burnt Common Nurseries near Send, 1970-3.
Tropical Bush-cricket *Phlugiolopsis henryi*	Colony in the Tropical Fern House at Kew Gardens in 1940s (specimen in R.H.S.Wisley Collection).
Tropical House Cricket *Gryllodes sigillatus*	Kew Gardens, 1976. Large colony in Houses 13D and 14D destroyed by fumigation.
Two-spotted Cricket *Gryllus bimaculatus*	A southern European cricket, very similar to the Field Cricket, now widely reared for the pet trade. A colony in shrubbery and soakaway grills by a petrol station at Wrythe Green, Carshalton, in 1995 (DAC).
Loboptera decipiens	A Mediterranean cockroach first recorded in Britain at Virginia Water in August, 1997. It was abundant in a house and garden and had to be destroyed because it can become a nuisance.
Surinam Cockroach *Pycnoscelus surinamensis*	Colony at Kew Gardens, 1954. Also at West Hall, Byfleet (22 specimens in R.H.S.Wisley Collection).

Common Cockroach *Blatta orientalis*	Possibly widespread. Found out of doors at Horley in 1983 (RDH).
American Cockroach *Periplaneta americana*	Kew, 1987 (AJP).
Australian Cockroach *Periplaneta australasiae*	Kew, 1987 (AJP). Colony for many years and still existing in heated greenhouses at R.H.S.Wisley.
German Cockroach *Blattella germanica*	Probably widespread in buildings. Common in flats at Clapham, 1989.
Brown-banded Cockroach *Supella longipalpa*	Chertsey, 1973. Croydon, 1978.
Ring-legged Earwig *Euboriella annulipes*	Kew, 1897. Abundant for a few years around 1970 at Callow Hill rubbish-tip near Egham until it was filled and landscaped (AJP).
Laboratory Stick-insect *Carausius morosus*	Established colony for many years in 1920s in greenhouses at J.Innes Horticultural Institute at Merton (specimens in R.H.S.Wisley Collection). Also for many years in 1950s in glasshouses at Kew Gardens.

There are also many other foreign species which are accidentally imported from time to time, mainly on fruit, and any of these could be found in Surrey. However none of these has been known to have established a breeding colony and they are treated as casual introductions.

SURREY — THE SURVEY AREA

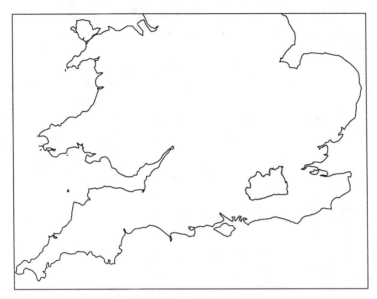

Surrey in relation to southern England

The boundary of the current county is a political boundary which has changed markedly in recent times and will no doubt change again, and is consequently wholly unsuited to the recording of any biological group over any period of time. For this a stable boundary is necessary and exists in the form of the vice-county, a division originally proposed by H.C. Watson in 1852 in order to provide a set of unit areas of roughly similar dimensions for botanical recording. The system was rapidly adopted by botanists and supported by the forerunners of the Botanical Society of the British Isles and has since been in use by many zoologists. Since the introduction of this system the political boundary of Surrey has changed several times and the sense of using the fixed boundary of the vice-county can immediately be seen. The whole vice-county system is explained in text and maps in Dandy (1969).

The vice-county of Surrey differs from the current administrative county principally in its northern boundary which is marked by the course of the River Thames thus including the boroughs of the south-western quadrant of Greater London. The southern boundary differs slightly in that it runs almost east-west in the vicinity of Horley and so includes the area occupied by Gatwick airport which is currently in West Sussex, and on the western boundary an area of approximately one square kilometre to the south of the village of Batt's Corner and part of present-day Surrey is excluded. The other major exclusion is the district of Spelthorne, which only became attached to Surrey in 1965, and in fact belongs principally to the vice-county of Middlesex. The actual boundary and that of the adjacent vice-counties is shown on the map below.

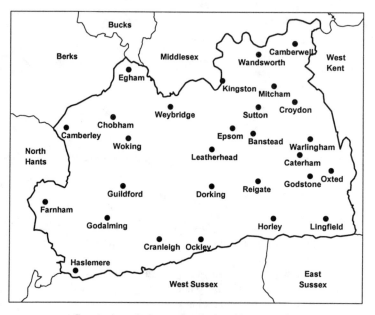

Surrey in relation to bordering vice-counties

CLIMATE, GEOLOGY AND DISTRIBUTION

Solid geology of Surrey, after **'Butterflies of Surrey'.**

Orthoptera, being a group of mainly warmth-loving insects, are on the edge of their range in the cool, damp and rather sunless climate of Britain, which explains why there are only 35 species here out of a total of roughly 20,000 species in the world. The distribution of these species within Britain and Surrey is dependent mainly on the climate, but also on geology and how this affects the habitat.

Surrey, in the extreme south-eastern corner of Britain, has a warmer and drier climate than most other counties and it is mainly for this reason that most of the British species are found here; the few species that are missing from Surrey are those which occur mainly on the coast or not far inland where the climate is even milder. Counties to the north of a line between the Wash and the Severn, with a much cooler climate, only have about half the number of species as Surrey.

Geology does also affect distribution; however it is not the actual soil types which have any great effect on distribution but rather the way in which geology affects the topography and thus the microclimate and habitat. Surrey is fortunate in having a varied geology, which has produced many hills with warm, south-facing slopes as well as extensive areas of poor sandy soils which have prevented any large-scale agriculture. Because of this the county still has a high proportion of semi-natural habitat, a feature which is preferred by most species of Orthoptera.

The basic solid geology of the county is fairly simple and is shown above. A ridge of chalk with a steep south-facing escarpment runs from west to east across the whole county, with a parallel ridge of greensand running to the south of it. In the west, on either side of these features, are large areas of sand: in the north the damper Bagshot sands overlying clay and in the south the drier Lower Greensand. In the east there are large areas of clay on either side of the chalk and sand ridges: in the north the acid London Clay and in the south the heavy but less acid Weald Clay. The best account of the geology of Surrey is by Stevens in Lousley (1976).

These varied soils produce equally varied habitat. Much of the North Downs is covered by thin chalk soils which are too dry for anything but grassland; the south-facing escarpment still has many areas of downland turf although this is rapidly disappearing due to lack of grazing. The hot, dry, sandy soils of the west support a rich heathland habitat but much of this has disappeared under encroaching pine and birch woodland, again due to lack of grazing. Much of the wet, heavy clay is still predominantly covered in oak woodland because it makes poor agricultural soil.

It is this variety of habitat which affects the distribution of some of the more localised species. The very local Woodland Grasshopper is almost restricted to the damp clay woodlands in the south, but only where there are open sunny glades or rides. The scarce Rufous Grasshopper is entirely restricted to grassland on warm, south-facing, chalk slopes of the North Downs, but never far from trees. The Mottled Grasshopper is almost restricted to dry heathland and the Bog Bush-cricket to damper heathland, although the former is also occasionally found on hot dry chalk slopes and the latter occurs less frequently on drier heathland and even in bilberry woods on the top of the highest hills in Surrey. The nationally rare Large Marsh Grasshopper has the most specialised habitat requirements of any British species; it only occurs on the wettest quaking sphagnum bogs on lowland heaths.

However many species are widespread and occur on different soils, so long as the habitat suits them. For instance the three commonest grasshoppers occur on all types of soil although they each favour a rather different habitat; the Common Green Grasshopper prefers rather long grass, the Field Grasshopper prefers short, dry, sparse grass and the Meadow Grasshopper prefers damper, lusher grass. The three common bush-crickets are also found on all soil types: the Oak Bush-cricket wherever there are suitable trees, the Speckled Bush-cricket and Dark Bush-cricket wherever there are suitable overgrown hedges and rough vegetation, although the latter avoids very dry situations on the sand and chalk.

The Stripe-winged Grasshopper has an interesting distribution in Surrey. It is normally regarded as an insect of short grassland on chalk or limestone but in Surrey it occurs just as frequently on acid grass heathland, especially sandy golf-courses, as on the southern slopes of the North Downs. It probably requires a habitat of warm, dry, well-drained soil regardless of actual soil type.

Two species of bush-cricket, Roesel's Bush-cricket and Long-winged Conehead, have recently invaded Surrey. They were previously entirely restricted in Britain to coastal areas because, being at the very northern limit of their range, they required a very warm climate. However with the recent series of warmer and longer summers they have been able to spread inland and are now found throughout the county on all types of soil. They are clearly more dependent on climate than geology.

THE FOSSIL RECORD
By Edmund A. Jarzembowski

The geology of the county is summarised in the companion volume *Dragonflies of Surrey*. Fossil Blattodea and Orthoptera are found in the Weald Clay of Early Cretaceous (Hauterivian-Barremian) age, some 121-132 million years Before Present in the latter part of the age of dinosaurs. New developments since 1996 include a detailed study of the Wealden palaeoclimate which is considered to have been of a 'Mediterranean' type (Allen, 1998). The Weald Clay in Surrey has been recognised by English Nature and the Countryside Commission as part of the Low Weald Natural and Character areas respectively.

Order Blattodea

Cockroaches are the second most abundant order in the Wealden after beetles (Jarzembowski, 1991). Fossils include body parts and wings (Figure 1), often with colour pattern preserved (Figure 2). The majority appear to belong to the Mesoblattinidae, an extinct Mesozoic family of ovipositor-bearing

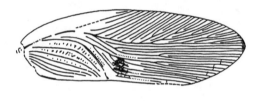

Figure 1. Forewing of *Elisama mollosus* Westwood, 1854 det. A. Ross. Upper Weald Clay, Auclaye Brickworks. (After Jarzembowski, 1987.) Scale 2mm.

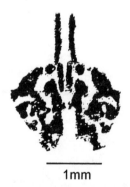

1mm

Figure 2. Pattern on *Gyna*-like pronotum. Smokejacks Brickworks. Imaged by Peter Green. MNEMG 1999.11 coll. Barbara Butler.

cockroaches. Cockroaches of more modern aspect have also been recognised such as a species resembling *Gyna* (Jarzembowski, 1987) and a member of the family Blattellidae (Vršanský, 1997). The occurrence of the latter is consistent with the record of a hymenopterous parasite of cockroaches (*Cretevania*: Jarzembowski, 1984). There are over 25 species of cockroaches in the English Lower Cretaceous which are currently being studied by Mr. A. Ross at the University of Brighton.

Order Dermaptera

Earwigs have not yet been recognised in the Wealden. This may be an oversight because of their resemblance to rove beetles.

Order Orthoptera

Grasshoppers, bush-crickets and crickets occur occasionally in the Wealden. They are diverse but it should be noted that grasses had not yet appeared in Cretaceous times. Bush-crickets (Ensifera) include Elcanidae (Figure 3; Plate 1, figure 1). The possible behaviour of this extinct Mesozoic family has been the subject of discussion, notably over the use of the moveable spines on the hindlegs e.g. Jarzembowski (1987); even a gregarious group habit has been postulated (Dr. J. Ansorge, *pers. comm.*). Large bush-crickets are less common and include extinct representatives of the now relict family Prophalangopsidae (Plate 1, figure 2) which is no longer found in Britain. The last fully-winged prophalangopsid was collected in India c.1861 but has not been recorded this century and is now feared extinct. The colour pattern may be preserved in fossil Prophalangopsidae, as in Elcanidae. Cricket finds include both sexes of Grylloidea. The presence of stridulatory files (Plate 2, figure 3) shows that extinct males could sing just as in Prophalangopsidae.

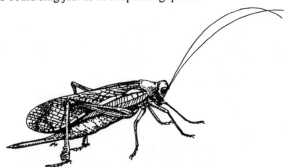

Figure 3. **Elcanid reconstructed by Neil Watson.**

Grasshoppers are rare. They include the extinct Mesozoic family Locustopseidae (Plate 2, figure 4) and an unusual acridoid of uncertain family (Figure 4). The Wealden Orthoptera are being studied in association with

Dr. A. Gorokhov (St. Petersburg). Moscow colleagues are investigating gut contents in more intact material from Asia; this is yielding evidence of diet in related Orthoptera of comparable age.

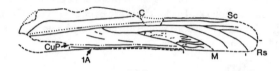

Figure 4. Acridoid forewing, Upper Weald Clay. Smokejacks Brickworks. (From Jarzembowski, 1987.) Scale 1mm.

This is PRIS Contribution no. 750.

REFERENCES

Allen, P., 1998.
Purbeck-Wealden (early Cretaceous) climates. *Proceedings of the Geologists' Association*, **109** (3): 197-236.

Jarzembowski, E.A., 1984.
Early Cretaceous insects from southern England. *Modern Geology*, **9**: 71-93, pls1-4.

Jarzembowski, E.A., 1987.
Early Cretaceous insects from southern England, 421 pp. PhD thesis, University of Reading.

Jarzembowski, E.A., 1991.
The Weald Clay of the Weald: report of 1988/89 field meetings. New insects from the Weald Clay of the Weald. *Proceedings of the Geologists' Association*, **102** (2): 83-108.

Jarzembowski, E.A., 1995.
Early Cretaceous insect faunas and palaeoenvironment. *Cretaceous Research*, **16** (6): 681-693.

Jarzembowski, E.A., & Coram, R., 1997.
New fossil insect records from the Purbeck of Dorset and the Wealden of the Weald. *Proceedings of the Dorset Natural History & Archaeological Society*, **118**: 119-124.

Vršanský, P. 1997.
Piniblattella gen. nov. – the most ancient genus of the family Blattellidae (Blattodea) from the Lower Cretaceous of Siberia. *Entomological Problems*, **28** (1): 67-79.

HISTORY OF RECORDING IN SURREY

Surrey is a county well-known not only for its Orthoptera but also for its resident orthopterists. The authors of three major works on the British Orthoptera, W.J.Lucas, D.R.Ragge and J.A.Marshall, were or are residents of Surrey. But in spite of this and although so close to London, Surrey was not a well-recorded county until this present survey was made. In fact the only published account of the Orthoptera of the whole county was that published in the *Victoria County History* (Burr, 1902); this was unfortunately brief and very general in its description of the species and concentrated almost more on which species had not been recorded than on those which had. Burr wrote: "The list of Orthoptera known to have been taken in Surrey is very meagre; this may be due to want of observation, for several good collecting grounds are within the borders of the county." Another reason was no doubt the lack of any book on Orthoptera at that time, although Burr himself had written a small book, published in 1897, called *British Orthoptera*; there was to be no other book until the publication in 1920 of Lucas's authoritative work *A Monograph of the British Orthoptera*.

However the first mention of Surrey in any work is probably as early as 1761, when on May 20th of that year the famous Gilbert White wrote in his *Naturalist's Journal* of the Field Cricket "abounding most in sand-banks on the sides of heaths, especially in Surrey and Sussex". Unfortunately this delightful insect became extinct in Surrey as recently as 1964.

Surrey is next mentioned in 1835 by J.F.Stephens in his *Illustrations of British Entomology* where he stated that the Mole-cricket was found at Ripley; this most elusive insect was recorded there regularly until 1953. The Mole-cricket is also mentioned in a little-known book published in about 1849 entitled *The Letters of Rusticus on Natural History* by Edward Newman; in this book it is described as abundant at Thursley Common, but it has not been recorded in the area for some years.

M.Burr's *British Orthoptera* of 1897 stated that the Rufous Grasshopper was found on Oxshott Common, but this was probably an error; Lucas doubted the record too. Although Burr at that time lived at East Grinstead he did not apparently know Surrey well, except around Dormans and Lingfield, and clearly accepted written records from many amateur entomologists who were not necessarily experienced in Orthoptera; in the preface to his later book *British Grasshoppers and Their Allies*, published in 1936, he refers to this earlier book which he describes "as a boyish effort, the first book on our British Orthoptera."

In October 1897 Lesne's Earwig was recorded in Surrey for the first time, when it was found on the North Downs near Reigate (Lucas, 1920).

In 1902 *The Victoria History of the County of Surrey* was published, the chapter on Orthoptera being written by Burr, with notes by W.J.Lucas. Most of the commoner and more widespread species had been found by then, but only a few localities were specifically mentioned. He said that the Woodland Grasshopper had been taken at Box Hill, but this was probably a case of mistaken identity as it has never been found there since. The House Cricket was still described as abundant. He did however give a list of species which, although not yet recorded at that time, he suggested should be sought for, and where to look for them, as follows:-

Short-winged Earwig ". . . because it has been found in Kent."

Lesser Cockroach ". . . which occurs more or less frequently in the southern counties."

Great Green Bush-cricket ". . . not actually recorded from within the county, but almost certain to occur"

Lesser Marsh Grasshopper ". . . on sandy heaths"

Large Marsh Grasshopper ". . . in marshy places"

Short-winged Cone-head ". . . rare and local, only among reeds in marshy places"

Grey Bush-cricket ". . . on chalk hills, especially among Rest Harrow . . . found in Kent"

Roesel's Bush-cricket ". . . in grassy fields . . . found in Kent"

Wart-biter ". . . on barren and arid spots, clearings in woods, etc . . . found in Kent and Hants"

Field Cricket ". . . on warm sandy commons"

Mole-cricket ". . . on warm sandy commons and in moist spots, where the soft ground permits it to make its well-known burrows."

All these species, apart from the Wart-biter and the Grey Bush-cricket, have since been found, although some only very recently. At that time it was not realised that the Grey Bush-cricket was restricted to the coast; it may now be spreading inland like the Long-winged Cone-head, and may be found in Surrey in the future. Most of those which have since been found are rare and local but others are now common and widespread. It is interesting that neither the Short-winged Cone-head nor the Lesser Marsh Grasshopper had been recorded by 1902, as they are both now quite widespread; it seems likely that they were overlooked and have since spread to become more common.

The period between 1902 and 1920 was a period of great activity amongst orthopterists. Lucas wrote regular articles and notes on Orthoptera, mainly in *The Entomologist*, continuing to contribute right into the 1930s. He lived first at Byfleet and then at Kingston upon Thames and he referred to many Surrey localities in his annual notes.

1920 saw the publication of the first major work on British Orthoptera, *A Monograph of the British Orthoptera* by Lucas. This book listed localities for each species by county and gave the first published modern Surrey records for five new species:-

Short-winged Cone-head . . . from Witley

Great Green Bush-cricket . . . from Godalming, Thursley and Ranmore

Lesser Marsh Grasshopper . . . from Tilford, Hale and Box Hill

Mole-cricket . . . from Churt (1901) and Milford (1902)

Field Cricket . . . from between Eashing and Godalming, and near Farnham

The Box Hill record for the Lesser Marsh Grasshopper was no doubt an error, as the habitat is not suitable. The Farnham record for the Field Cricket was taken from Burr, who had mistakenly placed Farnham in Hampshire; this locality may refer to the later well-known site between Frensham and Tilford, just south of Farnham.

Although Lucas's book must have encouraged entomologists to take a keener interest in Orthoptera, such interest must soon have waned as little new knowledge about them was gained for the next forty years or so. His book was not easy to use and the photographs were poor for identification purposes. Orthopterans never became as popular for study as other groups of insects because they were not very "collectable"; when stored in cabinets, they soon lost their colours and they then all looked rather similar and drab. Burr attempted to stimulate interest in recording species distribution with his little 1936 book *British Grasshoppers and Their Allies*, which for the first time included maps of Britain for each species; the distribution of each species was indicated by shading the counties (not vice-counties) in which it occurred.

However in 1944 Roesel's Bush-cricket was first found in Surrey at Cheam by Ian Menzies (although the specimen was originally misidentified as a Grey Bush-cricket). Then in 1958 R.M.Payne published a paper in *The London Naturalist* entitled "The distribution of grasshoppers and allied insects in the London Area", giving lists of localities for each species by county. It included many new localities in the north-east of Surrey and also a brief history of the discovery and subsequent spread of Roesel's Bush-cricket.

The first comprehensive survey of any part of Surrey was made by R.A.Farrow for the Croydon Natural History and Scientific Society and in 1962 he published "Orthoptera of the Survey Area" with concise species accounts and a distribution map, the Croydon survey area being much of the eastern part of the county.

At last, in 1965, a modern comprehensive book on Orthoptera was published; this was David Ragge's *Grasshoppers, Crickets and Cockroaches of the British Isles* and it contained not only excellent coloured illustrations, identification keys and distribution maps on a vice-county basis, but also song-diagrams. This book prompted a great new interest in the study of Orthoptera and in particular the recording of species distribution. The Biological Records Centre at Monks Wood started to produce Provisional Distribution Maps with Michael Skelton as editor; these maps of the British Isles were dot maps, prepared on a 10 km square basis, and they had the desired effect of encouraging amateurs to get out in the field and record the distribution of all species. This in turn led to the setting up of the National Mapping Scheme under Chris Haes and, in 1977, the issue of its first Newsletter; these Newsletters have continued to be issued annually (since 1996 under the new editor John Widgery).

It was as a result of these various new publications that I first became interested in Orthoptera in about 1967 and started to record sites where I had seen different species. I was greatly encouraged in this by Chris Haes, who not only helped me with identification but also took time to show me sites for the rarer species. At the time he was preparing a survey of the Sussex Orthoptera, which he later published (Haes, 1973, 1976), and encouraged me to do the same for Surrey; and so from 1970 onwards I kept notes of all localities for each species, with grid references. It soon became apparent to me that Surrey was just as rich a county for Orthoptera as Sussex, even though it had no coastline, but, operating as I was from my base in Milford, I was finding it difficult to record in the eastern part of the county. Luckily in 1976 Roger Hawkins arrived in Surrey and offered to start recording around his home in Horley, and from then on the draft distribution maps became much more balanced; a handful of other recorders also sent in records at this time. Roger Hawkins and I quickly learnt, by experience, different methods of recording large numbers of tetrads in a short space of time, and by 1980 the bulk of the mapping had been done; most of the remaining blank spaces had been filled in with the aid of bat-detectors by 1991. However at that time the Surrey Wildlife Trust was only just thinking about publishing the present series of atlases, and publication of the survey therefore had to wait.

The first definite Surrey record of the Lesser Cockroach came in 1969 when Valerie Brown found nymphs of this species at Chobham Common; there had been unsubstantiated reports from other places previously. In August 1982 the Large Marsh Grasshopper was added to the Surrey list, when T.Price recorded it from Folly Bog on West End Common.

The publication in 1988 of the next major work on British orthopteroids, *Grasshoppers and Allied Insects of Great Britain and Ireland* by Marshall and Haes, stimulated yet more interest among amateur orthopterists and as a result a few more people sent in their Surrey records.

In August 1990 Ian Menzies recorded the Long-winged Cone-head for the first time in Surrey, while trying out a bat-detector at Bookham Common; it was he who had found this insect for the first time in mainland Britain in 1945. It was found in two other places in Surrey a few days later and within five years it had become widespread and common throughout the county; at the same time Roesel's Bush-cricket was spreading out somewhat more slowly from its stronghold around London. This meant that the the whole county had to be resurveyed in order to monitor the extent and the speed of the spread of these two species; luckily by then we were all equipped with bat-detectors and the work could be done reasonably quickly, both species being easy to locate by this method..

Finally in September 1997 Roger Hawkins added the Short-winged Earwig to the Surrey list, when he discovered it while beating for other insects near Oxted. This brought the total number of native orthopteroids for Surrey to 30. Roger Hawkins also found Lesne's Earwig in 1997 on the North Downs and, rather surprisingly, quite commonly around Dunsfold on weald clay. This had always been considered to be a scarce insect, but because of these finds a last-minute survey of the North Downs was made in October 1998, in order to assess the distribution of this earwig; it was found to be common in most of the places searched. It had not originally been intended to include earwigs in this book, but as a result of these important finds it was decided at the last minute to include them.

IDENTIFICATION

The orthopteroids are a relatively small group of mainly large and mostly diurnal species; almost all of them, with practice, are quite easily identifiable in the field without the need to collect them. It may be necessary to catch some of the grasshoppers and turn them over to check a few details. All but two or three of the 30 Surrey species can be identified without much trouble in the course of one season. The group is comparable to that of the dragonflies, which has a similar number of species, all of which are diurnal and brightly coloured, making them easy to identify in the field. The problem with dragonflies is that they are great fliers, whereas the orthopterans are much more static and therefore easier to examine and to catch if necessary. But orthopterists have one major advantage over odonatists when recording or surveying; all the grasshoppers and bush-crickets sing and the song of each species is so distinctive that they can be quite easily identified by song alone, without even seeing them. This is particularly helpful with those species of bush-cricket which sing at night. In fact the songs of the grasshoppers are more reliable as a means of identification than views of the species in the field. The different songs are dealt with in detail in the next chapter.

The bush-crickets are the easiest group to identify, since they all have easily seen distinctive characters; even the nymphs are readily identifiable, except for the two cone-heads. The grasshoppers are rather more difficult, partly because each species may vary so much in its colour pattern; colour is not always a useful distinguishing character in many grasshoppers and other characters must be used. The smaller nymphs are almost impossible to identify. The cockroaches and earwigs are more difficult still, and some of these do need to be collected and examined at home, especially nymphs.

It is hoped that the colour-plates in this book will help readers to identify some species but they are not intended to solve all problems of identification. Luckily there are two books dealing with identification which are readily available, and a third which is now sadly out of print. The latter is David Ragge's *Grasshoppers, Crickets and Cockroaches of the British Isles*, which is very comprehensive and has excellent coloured illustrations, keys for identification and song-diagrams (these song-diagrams are reproduced in the next chapter). Another very comprehensive book is Marshall and Haes' *Grasshoppers and Allied Insects of Great Britain and Ireland*, which has excellent keys and coloured illustrations, and also dot maps showing the distribution of each species, although these maps are now rather out of date for a few species, even in the ten years since publication. The third book, Collins *Field Guide to the Grasshoppers and Crickets of Britain and Northern Europe* by H. Bellmann, is a much smaller pocket-book with very good coloured photographs of each species, keys for identification and song-diagrams, but it covers all 78 central European species, of which less than half are British.

In addition to these major works, there are two smaller but useful booklets. The first is the handbook, *Grasshoppers* by Valerie Brown, in the Naturalists' Handbooks series, with excellent coloured illustrations and identification keys to the grasshoppers and bush-crickets, and the second is the booklet in the Shire Natural History series, *Grasshoppers and Bush-crickets of the British Isles* by A. Mahon, containing song patterns and tips on identification. There are also two articles on identification of grasshoppers and crickets by Paul Sterry in Vols. 1 and 2 of *British Wildlife*; this magazine has regular features on Orthoptera.

The keys for identification in these books will help readers to pick out the salient characters which distinguish one species from another similar species. Once these characters have been learnt, identification in the field is reasonably straightforward. But the best way of learning these field characters is to go out in the field with an experienced orthopterist (preferably also a good field naturalist) who will know where to go and how to find each species and will be able to point out all the distinguishing characters, as well as the habits and songs.

SONGS AND BAT-DETECTORS

The crickets and grasshoppers are best known for their songs, which are almost unique in the insect world, the cicadas being the only other group that produce comparable sounds. In fact the word "cricket" is derived from the Old French onomatopoeic word "criquer" meaning "to crackle". All the British species of bush-crickets, crickets, mole-cricket and grasshoppers produce songs using various techniques; the only groups which are silent are the groundhoppers, the cockroaches and the earwigs.

The different techniques for producing sounds produce in turn a very varied range of songs, varying in pattern, pitch and volume; the Great Green Bush-cricket has perhaps the loudest song of any insect found in Britain and can be heard from about 50 metres, whereas the Oak Bush-cricket makes such a soft noise that it can only just be heard by a few humans at one metre.

In grasshoppers the normal technique is to rub the hind femora, on each of which is a row of stridulatory pegs, up and down on a prominent vein of the forewings; the noise produced by different species varies with the number and spacing of the pegs as well as the speed of the hind leg movements. The Large Marsh Grasshopper is almost unique among grasshoppers in that it kicks its hind leg against its forewing producing a loud and distinctive clicking sound. In most species it is normally the males that sing but the females can and do occasionally sing, especially when unmated and in close proximity to a singing male.

Bush-crickets use an entirely different technique which in turn produces a very different kind of sound. They raise their forewings slightly and then rub the left one, on which is a small stridulatory ridge with teeth, over a scraper on the right one, causing the wings to vibrate. The noise produced varies between species mainly in volume and pitch but also in spacing. The Oak Bush-cricket is unique in using not its wings but its feet for its song; it drums with its hind leg on a leaf, thus producing a very soft pattering sound, inaudible to most humans. Only the male bush-crickets sing, although female Speckled Bush-crickets are now known to sing quietly in response to the males.

Crickets vibrate their wings in a similar way to bush-crickets, but a stridulatory ridge on the right forewing is rubbed over a scraper on the left one; their forewings also have two large membranous surfaces, called the harp and the mirror, which amplify the sound (the bush-crickets only having the mirror). The Mole-cricket uses the same technique as the crickets, but it sings from a specially constructed, resonating, singing chamber, just below the surface of the soil, with two or three holes leading to the surface, from which the sound carries for a considerable distance. Because the song resonates it is louder and more audible than the songs of other crickets.

Because each species produces its own distinct song it is perfectly possible to identify all the British sound-producing orthopterans solely by their songs, in just the same way that most birds can be identified by their songs. Some species are in fact easier to identify by their songs than by their appearance and structural characters, e.g. the separation of the Long-winged Cone-head from the long-winged form of the Short-winged Cone-head f.*burri*. Also some species are notoriously difficult to find by sight but can easily be located by sound from some distance, and identified by their song, e.g. the cone-heads. When surveying an area for Orthoptera, a recorder can save an enormous amount of time if he can identify the different species by their songs without the need to find and catch or see each one; many tetrads can be covered in a short time.

The songs are so varied in shape or pattern, in spacing, in volume, in range of frequency and in pitch that they are difficult to describe adequately in words; the same problem arises in describing birdsong. Any such description is necessarily subjective because each person has his own idea of what the song sounds like. Another problem is that many of the songs, particularly those of the bush-crickets, are at a very high frequency and as a result are inaudible to many adults, especially older ones. However these problems can be overcome, if the songs are not only described in words, but are also illustrated by oscillograms or song-diagrams.

The best way of learning the different songs is to listen to high quality tape-recordings of these songs and luckily these are now available on cassette and on compact disc. *A Sound Guide to the Grasshoppers and Allied Insects of Great Britain and Ireland*, by J.F.Burton and D.R.Ragge, is a 30 minute cassette of the songs of 26 species, and *A Sound Guide to the Grasshoppers and Crickets of Western Europe*, by D.R.Ragge and W.J.Reynolds, is a set of two CDs of the songs of 170 species, including all those found in Britain.

Yet another very useful tool for identifying the songs of grasshoppers, and particularly bush-crickets, is the ultrasound detector or mini bat-detector, which converts the frequencies that are normally too high for human hearing into easily audible sounds. Not only does this tool make it easy to hear the pattern or cadence of the song, but it also has a frequency dial on it, which clearly shows the frequency on which the particular species is singing. This can be extremely helpful, because similar-sounding and similar-looking species can be distinguished by the different frequencies shown on the dial. Also the song can be picked up on a bat-detector from a far greater distance than from which it can be heard by the human ear. The bat-detector can be fitted with a directional horn and the recorder can then pinpoint the position of the cricket and, with luck and persistence, see it. Bat-detectors (modified to a lower frequency to pick up grasshoppers on 10 kHz) can be obtained from Ultra Sound Advice of 23 Aberdeen Road, London N5 2UG (tel. 0171 359 1718).

Table 1 below shows the song-diagrams of the Surrey species, which are reproduced with the consent of the author David R. Ragge from his book *Grasshoppers, Crickets and Cockroaches of the British Isles*. Each song is represented as a horizontal pattern arranged on a time scale in seconds. Each black line or mark denotes a sound and the spaces between them represent silence. Where no spaces are shown the sound is continuous or at least appears so to the human ear. As can be seen most of the songs are composed of a series of distinct sounds, or chirps, represented by the vertical black marks. The vertical spread of each horizontal song-diagram is intended to give a very rough measure of the relative loudness of the sound. Thus the song-diagram of the Great Green Bush-cricket has a broader vertical spread than that of the Oak Bush-cricket, which produces a much fainter sound, and the song of the Common Green Grasshopper is shown to reach its maximum loudness about half-way through its duration. It may be useful for readers to listen to a recording of the songs whilst comparing the song-diagrams. If older readers have difficulty in hearing the recordings of the high frequency songs of the bush-crickets, the use of a bat-detector pointed at the speaker will overcome this difficulty and will also indicate the frequency of the song. However it should be noted that there is no ultrasound on the cassette and CDs, so that the bat-detector will not give a true indication of the dominant frequency.

TABLE 1

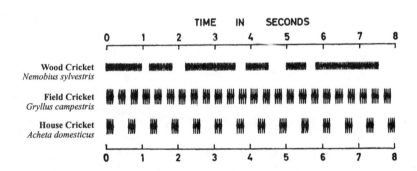

Song-diagrams of the Surrey crickets

GRASSHOPPERS AND CRICKETS OF SURREY – INTRODUCTION

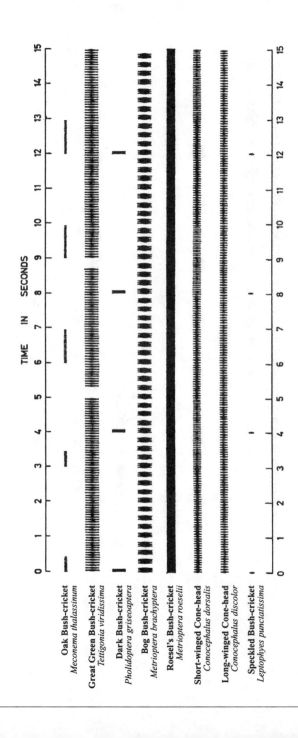

Song-diagrams of the Surrey bush-crickets

GRASSHOPPERS AND CRICKETS OF SURREY – INTRODUCTION

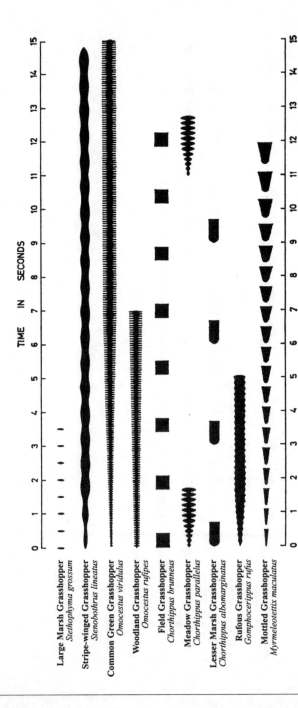

Song-diagrams of the Surrey grasshoppers

Table 2 overleaf gives a brief verbal description of the songs of each of the Surrey species of grasshoppers and bush-crickets, and also the measurement of peak frequency. It will be seen that all the grasshoppers sing on a much lower frequency than bush-crickets (due mainly to the larger membrane on the wings of grasshoppers, but other factors are involved and are sometimes more important) and all on much the same peak frequency of about 10 kHz (except the Meadow Grasshopper which can be picked up on a bat-detector on 22 kHz as well as on 11 kHz). Grasshoppers can be difficult to hear on such a low frequency because the bat-detector will also pick up other background noises, such as the rustling of grass-stems if there is the slightest wind. However, because of this low frequency, many people can still hear the songs of grasshoppers even when the songs of bush-crickets, especially coneheads, have become inaudible to them. Mike Thurner produced a similar table in an article entitled 'Use of a bat detector for Orthoptera surveys', published in *Surrey Biological Recorders Newsletter* No 3, November 1994.

Most of the bush-crickets call on a very high frequency and hence are inaudible to many people, especially older ones, unless they are lucky enough to have acute hearing. But these calls are very easily heard on a bat-detector, even at long distances; for instance the Speckled Bush-cricket can normally be heard only by people with a very acute sense of hearing from about 1 metre, but on a bat-detector it can be heard clearly from about 30 metres. With these species the use of a bat-detector is essential. With some species the use of a bat-detector is helpful to those with poor hearing, and increases the distance from which they can be heard, even for those with good hearing; Roesel's Bush cricket is one of these and the noise of its call on the bat-detector is so loud that it can drown out the calls of other species. Other species may not be heard better, or even not be heard at all with a bat-detector; all the crickets come into this category. In the Table a note is made if a bat-detector is essential, helpful or not helpful. A note is also made in the Table, where appropriate, of the approximate distances from which people with good high frequency hearing can normally hear the songs, first without (w/o) a bat-detector and secondly with a bat-detector. Some species call over a broad or even very broad frequency spectrum and it is then necessary to turn the frequency dial on the bat-detector until the user finds the dominant frequency. In such cases the frequencies given in the Table are the dominant ones, although these can be subjective, as are the verbal descriptions.

TABLE 2

BUSH-CRICKETS

Great Green Bush-cricket ... Detector not helpful .. w/o 45m, with 30m .... 10 kHz
Very loud, continuous, rattling call, with short pauses

Dark Bush-cricket .... Detector helpful .... 22 kHz
Loud, slurred click every 4 seconds or so

Bog Bush-cricket. .... Detector helpful .... w/o 5m, with 20m ..... 25 kHz
Regular, continuous chuffing, like a steam train

Roesel's Bush-cricket .... Detector helpful .... w/o 10m, with 25m ... 20 kHz
Very loud, high-pitched, continuous buzzing like crackle from overhead electricity lines

Long-winged Cone-head ...... Detector essential ..... w/o 2m, with 25m ..... 30 kHz
Faint, high-pitched, continuous, buzzing chirp at constant speed

Short-winged Cone-head ..... Detector essential ..... w/o 1m, with 25m ..... 40 kHz
Faint, continuous chirp, alternating in speed; buzzing chirp and then slower ticking

Speckled Bush-cricket .... Detector essential ..... w/o1m, with 30m ...... 40 kHz
Faint, high-pitched, short tick at irregular intervals of about 5 seconds

GRASSHOPPERS

Large Marsh Grasshopper .... 10 kHz
Loud tick, repeated slowly about 8 times

Stripe-winged Grasshopper .... 10 kHz
Faint, high-pitched, slowly pulsating song lasting about 15 seconds

Woodland Grasshopper .... 10 kHz
Short, continuous trill of about 5 seconds, increasing in volume but stopping suddenly

Common Green Grasshopper .... 10 kHz
Similar to Woodland Grasshopper but a longer, continuous trill of up to 15 seconds

Field Grasshopper .... 10 kHz
Series of intermittent, hissing chirps

Meadow Grasshopper .... Detector helpful .... w/o 5m with 10m ...... 11 kHz
Short burst of pulsating chirps, repeated every 10 seconds   and 22 kHz

Lesser Marsh Grasshopper .... 10 kHz
Series of short soft chirps, about 3 seconds apart

Rufous Grasshopper .... 10 kHz
Soft, buzzing chirp for 5 seconds

Mottled Grasshopper .... 10 kHz
Long series of short buzzing chirps, starting softly but slowly increasing in volume

EXPLANATION OF SPECIES ACCOUNTS AND DISTRIBUTION MAPS

The accounts contain details of all the native species recorded in Surrey. The main aim of this book is to establish the present status and distribution of each species in Surrey and for this a period of 29 years, the survey period, has been used.

Each account starts with a brief description of the species, picking out the main characteristics which distinguish it from similar species. The colour plates should also assist in identification. However this book does not attempt to be an identification guide and readers should consult the books mentioned in the earlier chapter on Identification.

The description is followed by a few brief notes on the life history and habits of the species, together with habitat preferences, and a short description of the song or stridulation. Songs have been dealt with more fully in the previous chapter. The account then briefly mentions the status and distribution in the British Isles, but these notes are not meant to be exhaustive and readers should again consult Marshall and Haes (1988) and also the more recent *Atlas of Grasshoppers, Crickets and Allied Insects in Britain and Ireland* (Haes and Harding, 1997). Because they are both now critically endangered species in Britain, and either still occur or once occurred in Surrey, the accounts of the Field Cricket and Mole-cricket are much more extensive and give all the known Surrey data, both historical and current.

If a species is classified as having a national status of Nationally Scarce or RDB (*Red Data Book*), this is mentioned. However the Nationally Scarce status is only provisional; two Nationally Scarce species have spread so rapidly in the last decade that they are now widespread and abundant over a large area of southern England. It must also be stressed that it is a national status and not a Surrey status; there are many species which are Nationally Scarce but which are widespread and abundant in parts of Surrey, e.g. the Bog Bush-cricket, the Rufous Grasshopper and the Dusky Cockroach. The classification is made according to the number of 10 km squares in Great Britain from which the species has been recorded since 1980, as follows:-

Nationally Scarce   Species recorded from more than 15 squares, but not more than 100, subdivided into Nationally Scarce A (31-100 squares) and Nationally Scarce B (16-30 squares).

RDB   Red Data Book species are those recorded for 15 or fewer squares and are divided into classes based on degree of threat: RDB3, rare; RDB2, vulnerable; RDB1, endangered.

The account then deals with the past and current status and distribution of the species in Surrey, with notes of any interesting habits, habitat preferences, the best method of finding the species, and late dates which have been recorded in Surrey. When commenting on status, extensive reference is made to older authors (mainly Burr, 1902, and Lucas, 1920) wherever relevant.

Each species account (except the Common Earwig), is accompanied by a distribution map showing from which tetrads that species has been recorded, a tetrad being an area 2 km square. The national grid lines are shown, allowing the location of sites using the grid references given in the gazetteer, as are the boundaries of the principal geological formations, and the boundaries of adjoining vice-counties.

A solid dot represents a record made during the survey period, i.e. since 1970. A shaded dot represents a record made prior to 1970. In the case of the two species which have spread recently, Long-winged Cone-head and Roesel's Bush-cricket, extra maps are provided in order to show how the spread occurred.

FURTHER READING

PUBLISHED SURREY LISTS

Burr, M.D., 1902. with notes by W.J.Lucas.
Orthoptera in *A History of the County of Surrey*, 3 Zoology. Constable, London. (Victoria County History)

Lucas, W.J., 1920.
A Monograph of the British Orthoptera. London: Ray Society.

Payne, R.M., 1958.
The Distribution of Grasshoppers and allied insects in the London area. *London Naturalist* **37**: 102-115.

Farrow, R.A., 1962.
Orthoptera of the Survey Area (Grasshoppers and Crickets). Croydon Natural History and Scientific Society. Regional Survey, Index No.51.

Paul, J., 1995.
Orthoptera in the London Archipelago. *Entomologist's Record* **107**: 89-95.

CURRENT PUBLISHED LISTS FOR SOUTH-EASTERN COUNTIES

ESSEX

Wake, A.J., 1997.
Grasshoppers and Crickets (Orthoptera) of Essex.
Colchester: Colchester Natural History Society.

SUSSEX

Haes, E.C.M., 1976.
Orthoptera in Sussex. *Entomologist's Gazette* **27**: 181-202.

BERKSHIRE, BUCKINGHAMSHIRE AND OXFORDSHIRE

Paul, J., 1989.
Grasshoppers and Crickets of Berkshire, Buckinghamshire and Oxfordshire. Oxford: Pisces.

HERTFORDSHIRE

Widgery, J.P., 1991.
A provisional commentary on the status of crickets, grasshoppers and related insects (Orthopteroids) in Hertfordshire. *Transactions of the Hertfordshire Natural History Society* **31**:18-24.

FUTURE RECORDING

Recording the distribution of Surrey Orthoptera will not cease at the end of this survey; each species map is merely a snapshot taken at a certain period of time showing the range of that species at that time. In fifty or even in ten years' time that species map could look very different. Some species are always on the move, their range contracting or increasing as a result of climatic change, loss of habitat, agricultural methods, human pressure or a host of other factors. Roesel's Bush-cricket, for instance, only just had a toe-hold in the north-east corner of the county at the start of the survey, but by the end it had spread over the whole county; this spread was probably helped by a series of hot summers and changes in agriculture such as "set-aside". But who in 1970 would have dreamt that the Long-winged Cone-head would ever be recorded in Surrey, let alone that in a matter of a few years it would become one of the commonest bush-crickets throughout the county?

If global warming is a reality, then with hotter summers and longer, warmer autumns almost anything could happen. Already the strictly coastal Grey Bush-cricket has started to spread inland in the last year or two, and it could possibly repeat the invasion of Surrey by Roesel's Bush-cricket; in Normandy, France, it is just as common inland as on the coast. Until recently Cepero's Groundhopper was considered to be a coastal species, but it is now occasionally turning up at inland sand-pits; it may have been overlooked because of its close similarity to the Slender Groundhopper. And even the rare and very local Heath Grasshopper, which is restricted to east Dorset and the New Forest area, has very recently been reported to be spreading; there is plenty of suitable dry heathland habitat in west Surrey if it spreads eastwards.

But the species which are already resident in Surrey need to be monitored from time to time; for instance the Stripe-winged Grasshopper may well be decreasing because short downland turf is disappearing with the lack of grazing. Other species, such as the Woodland Grasshopper and the Dusky Cockroach, have restricted ranges but are common within those ranges. Neither is common at the edge of its range but they may possibly occur in discrete colonies outside their normal ranges; the Woodland Grasshopper might be found in the north-west of the county, since it occurs just over the boundary in Windsor Great Park. Other species, such as the Wood Cricket, may be increasing and spreading; a careful check of the Wood Cricket's range should be carried out to see if it is in woodlands surrounding Wisley Common or even further away. The two species which appear to have increased their ranges during the survey period, the Short-winged Cone-head and the Lesser Marsh Grasshopper, need to be monitored to see whether their spread continues. In the case of the two species which have spread rapidly across the whole county, Roesel's Bush-cricket and the Long-winged Cone-head, further survey work needs to be carried out in order to see whether they eventually extend to every tetrad or whether they decline.

The native cockroaches and especially the earwigs, being small and difficult to find, are under-recorded and indeed many orthopterists do not bother to record them because of difficulties in identification. However it is hoped that the distribution maps in this book, coupled with the colour photographs and hints on how to find them, will encourage recorders to take an interest in these neglected groups. Lesne's Earwig may well be found in other areas if it is systematically searched for by beating, and the very localised Short-winged Earwig may prove to have a rather wider range than is shown on the map. The Lesser Cockroach is notoriously elusive but there is no shortage of dry heathland in west Surrey where a colony could be located, perhaps by searching for nymphs on buttercup flowers.

A careful watch will need to be kept on the few Surrey rarities to ensure that they and their habitat are conserved. The Great Green Bush-cricket is quite widely dispersed in its only two known sites but it could disappear from either site if there was a drastic change in the management of the area, and neither site as yet benefits from any statutory or voluntary control. In spite of its great size, it is a difficult insect to see and it could still be found at other sites, especially in warm places on the North Downs. The only known native locality for the spectacular Large Marsh Grasshopper needs to be more thoroughly investigated, but unfortunately almost all the best-looking habitat is within a military firing range, and recorders will be forced to search other nearby bogs with the hope of finding another colony. The bog at Thursley Common needs to be searched thoroughly in hot weather to see if the introduced colony still survives.

Finally to the Field Cricket and the Mole-cricket; if the reintroduction of the former proves to be successful, any colonies will need to be carefully observed. But the greatest prize of all will go to the recorder who eventually refinds a colony of Mole-crickets after many years of searching. There is a strong possibility that it still survives in the valley of the River Wey.

Readers should send all records to me for the time being and I will then update the Distribution Maps, with a view to publishing revised maps in due course. I will pass on any records of national interest to the National Orthoptera Recorder. However I would like to hand over the post of County Recorder to anyone who is willing to take over.

TETTIGONIIDAE – Bush-crickets including Cone-heads

The bush-crickets are distinguished from the grasshoppers by their extremely long, thread-like antennae and long and slender hind-legs. Not all species have fully-developed wings but they are very agile. The females have distinctive long curved ovipositors, used for depositing eggs in plant stems, cracks in bark or in the soil, and the males have characteristic songs which they make by rubbing their forewings together (stridulating). The two cone-heads are rather smaller and more slender; they are equally agile but are found on rushes and grass stems. It seems that most species are omnivorous but the Oak Bush-cricket is carnivorous, eating various insects. There are ten species of bush-cricket in Britain and only two of them, one very rare and the other coastal, are absent from Surrey.

Meconema thalassinum (De Geer, 1773)   PLATE 3   Oak Bush-cricket

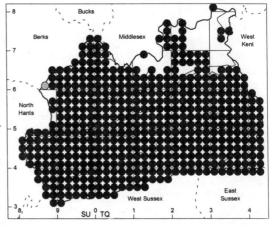

The Oak Bush-cricket is a small fragile-looking cricket, with fully developed wings, which flies short distances at night and often comes to lighted windows in the autumn. It is a most distinctive pale bluish-green in colour (or sea green, hence its latinised Greek name *thalassinum*) quite unlike that of any other British orthopteran. The female has a long green ovipositor which she uses to lay her eggs in crevices in the bark of deciduous trees. It makes no song but the male drums with its hind legs on leaves, making a faint pattering noise, scarcely audible to humans especially in outdoor conditions. It is an arboreal and nocturnal cricket, spending the day on the undersides of leaves. It is widely distributed over most of Britain and very common in the south.

Because it is entirely arboreal, nocturnal and effectively silent this bush-cricket can be difficult to locate. However it does fly freely and many records in the past have been of specimens coming to lights in houses or moth-traps. For this reason the coverage of records nationally is still patchy although it is probably common and widespread in southern England except in treeless areas. It is found mostly on oak and lime but also occasionally on poplar, maple, beech, ash, hawthorn and hazel. It does not appear to occur on plane trees which may explain why it is very scarce in Inner London. It has recently been recorded on soft-leaved conifers in suburban localities.

Although our records include specimens attracted to the light of moth traps, or coming indoors to light, or attracted to the window of a room lit by a strip-light, along with moths and lace-wings, rather more records refer to adult insects coming indoors through an open window into an unlit room, often during rain or cool weather. On a tree-trunk, the insects are disturbed by the sudden glare of a torch and often retreat behind the tree or even jump off.

Roger Hawkins has made numerous observations on the life habits of this species whilst recording them at night. From 1979 onwards he travelled around east Surrey on warm September evenings, using a torch to look for them on the trunks of trees. They were very easy to find, occurring on most large, isolated trees. Within woods, where there are more small trunks, each with a limited area of canopy foliage, they were less frequent.

The insects appeared on the trunks about half-an-hour after sunset. They probably walk down, since they can move very quickly over the trunks. Examining the branches of an oak tree at dusk once revealed at least three males walking rapidly, almost running, above or under the twigs. They may possibly sometimes fly down, but he has not observed this. Males are very evident early in the night (and early in the season), but later only females are seen. These move rather slowly when examining cracks in the bark and searching for a suitable egg-laying site. He has seen females laying eggs in bark crevices on over a hundred separate occasions. Oak and horse-chestnut trees are most favoured, but they also frequently lay on ash, sycamore, lime, sweet chestnut, poplar, false-acacia and beech, and occasionally on birch, Norway maple, hornbeam and ornamental cherry. More unusual sites include the trunk of a Lawson's cypress, a burnt stump and even a telegraph pole. When laying on smooth-barked trees such as beech, he has seen them choose a rotten bark-less part of the trunk, or moss at the foot of the tree. In Horley on 29 August 1984, four separate females were laying into a sheet of *Dicranoweisia* moss on the north side of a single ash tree of no great size. Egg-laying sites observed have been from close to ground level to 4 metres up, and probably higher. They continue laying in light drizzle, but tend to keep to dry parts of the trunk. In warm weather they go on laying at least until midnight. If it becomes cold (below about 7°C), the insects stop laying and move slowly back up the trunk.

The nymphs arising from all these eggs must at some time in the spring hatch and climb up the trees in order to spend their first day of life camouflaged under a leaf. A tiny nymph found alive but trapped in a spider's web, at 7 p.m. on 26 May 1981 at 2 metres up on a horse-chestnut trunk in Horley, was probably newly-hatched. Nymphs are more usually seen during the day-time under the leaves of shrubs or the lower branches of trees, or by beating from such situations. They are found especially on sallow bushes, but even on one occasion on box at Box Hill. When undisturbed, both nymphs and adults are invariably found under a leaf during the day.

As will be seen from the map, the Oak Bush-cricket is widespread throughout Surrey except in the most heavily built-up areas in the north-east, which is probably due to lack of suitable trees. It occurs for instance quite commonly at Kew, Richmond Park and Wimbledon Common which are all well-wooded, but it proved difficult to find any nearer to central London, in spite of numerous searches, until after a gale in mid-October 1991 when one was eventually found squashed under old oak-trees on College Road, Dulwich, where it adjoins Dulwich Wood (DWB). It is probably also present in the tetrads to the south of Dulwich. In 1994 it was found even further into Central London at Ruskin Park and Denmark Hill in Camberwell, where it was on limes and horse-chestnuts (Paul, 1995). It is also found regularly in Nunhead Cemetery, Peckham (RAJ). In August 1997 it was even found outside the National Theatre on South Bank; one had flown onto a car windscreen but, as there were no trees nearby, it may have been brought in on a vehicle unless it had flown some distance (RAJ).

All the old authors described it as common and widespread, although none of them knew of it nearer to London than Kew Gardens. It is still just as common now, which is hardly surprising as Surrey is among the most wooded of all counties and it may be commoner in Surrey than anywhere else.

Apart from finding stray specimens in houses, attracted by lights, or in moth-traps, the usually recommended method of finding this bush-cricket was to beat the lower branches of trees or tall shrubs. This was a rather laborious, time-consuming and mostly unproductive method and by the mid-seventies it had only produced records in a few tetrads. Another recommendation was to search the underside of leaves on oak trees using binoculars, but this proved to be even less productive. Then in the late seventies I discovered a method which revolutionised my speed of recording. I noticed that when squashed on a road by cars, this bush-cricket was very easy to see due to its unique bluish-green colour and I could even see them whilst driving slowly along country lanes. The squashed bodies remain perfectly recognisable on quiet roads for up to a week, so long as there has not been heavy rain to wash them away. They are most easily found, after a gale in the autumn, under oak or lime trees on less busy roads. It is possible to find them on busy roads but recording them can then become a dangerous sport! Occasionally Speckled Bush-crickets can be found by this method but their colour has no blue in it; confusion is also possible with squashed caterpillars. We refer to this as the "squashed on the road" method.

From 1979 Roger Hawkins in the south-east discovered that he too could record it in nearly every tetrad by using an entirely different method, which was visiting oak trees at night with a torch and finding the insects on trunks, and especially the females ovipositing in cracks of the bark at various heights above ground. Once again, because of their distinctive blue-green colour, they were very easy to find in the torch-light. Before 1979 he had only made two records in three years, but by using this method he was able to add 12 tetrad records in one evening while cycling over the flat Surrey Weald. We refer to this as the "torch-light" method.

Almost all the records from the west are mine using the "squashed on the road" method, whereas almost all those from the east are of Roger Hawkins using the "torch-light" method. If other orthopterists used these two methods, the National Atlas would undoubtedly have far more records, and the species might well be recorded even in new areas in the north of England.

Late Dates. This is one of the latest bush-crickets to mature. One was in a house at Milford on 6 November 1978 (DWB) and a female was still ovipositing on 11 November 1984 at Horley (RDH).

Tettigonia viridissima (Linnaeus, 1758)   PLATE 3   Great Green Bush-cricket

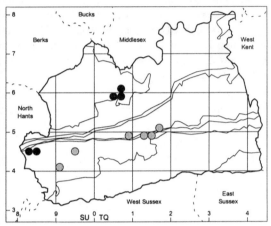

This is one of the largest and most impressive of British insects, being over 4 cm long and bright green all over (except for a brown stripe on its back) and with large green wings. Although easily identifiable it is not at all easy to find, unless one hears the high-pitched but very loud, continuous and far-reaching stridulation of the male which it makes from late afternoon into the night in late July and August. It prefers tangled overgrown vegetation and blends into the surrounding habitat remarkably well. The nymphs are often easier to find in early summer since they usually frequent rough grass at this stage and are more numerous and more easily disturbed than adults. They often occur in gardens, overgrown hedges and bramble patches.

In Britain it is confined mainly to southern coastal areas with a few scattered inland localities. It has only been known from two small areas in Surrey during the survey period but in view of our experience of overlooking one of these for 20 years, there is a real possibility that it could still be at other sites yet to be discovered.

Stephens (1835) recorded it from Battersea Fields, but Burr (1902) said that it had not been recorded in the county. O.H.Latter recorded it from Godalming in 1902. Lucas (1920) reported this, and others from near Thursley and at Pickett's Hole, Ranmore; in 1919 he also reported one being found at Westcott in 1918. Pickett's Hole is a steep south-facing slope on the North Downs just above Westcott. In 1935 J.L.Harrison recorded it from Hackhurst Downs about 2 km to the west. In August 1960 Chris Haes heard one stridulating at Westcott Green and in September 1961 C.J.Hodgson saw one at the same place. No records have been received from this locality since then, or from any of its former sites along the North Downs, in spite of several searches.

There has been a well-known colony at Middle Bourne, Farnham, since at least the 1950s. I was shown my first Surrey Great Green Bush-cricket there by Chris Haes when we saw five nymphs in rough grass in Middle Bourne Lane on 5 June 1974, and numbers have been seen or heard by various recorders in a large area along the valley of the Bourne from then up to 1997. This locality is quite built-up but there are large scrubby gardens on the south-facing slope of the little stream and the colony extends for more than one kilometre, both east and west of the Tilford road. In the mid-1980s the largest concentration was in the area around the Fox Inn on the Tilford road where Howard Inns could sit in the garden drinking his beer on a warm evening in August listening to a chorus of some ten singing males; he normally hears singing males from his garden in Lower Bourne by late July. Friends of his, who lived opposite the pub until 1987, complained of being kept awake at night by the crickets singing in their garden. In 1992 the stream was "tidied up and channelised"; the vegetation on the banks above was removed and the banks graded, which had a significant effect on the species, and since then the colony seems to have spread out and become significantly weaker. In 1994 Howard Inns was only able to count 12 singing males along the whole of the Bourne Stream. In 1997 a friendly neighbour presented him with an undamaged male which had been caught by his cat.

The other colony at Wisley has such a strange story that it is worth relating in detail. It was apparently unrecorded until 1991 in which year I was told by A.R.G.Mundell that he had won a Great Green Bush-cricket in a jam-jar as a prize at an Alpine Society meeting in Woking in 1984. It had come from R.Haywood who lived at Wisley and who had found it on runner beans in his garden. When I telephoned his widow in 1991, she told me that they had always had them in the garden and they made such a noise at night that they kept her awake. She had seen three in her garden in 1991 but had no idea that they were uncommon. Andrew Halstead, the entomologist at R.H.S.Wisley, told me that he had been given a photograph of one, taken in another garden in Wisley village in 1988. It seemed clear that there had been a thriving colony at Wisley for some years and yet it had gone unreported.

In 1992 I visited Wisley village and found the bush-cricket calling from an overgrown garden but failed to find it elsewhere. However I then started to get reports of people hearing its call whilst driving along the M25 at Wisley. It is well-known that the call, being so loud, can be heard from a moving car at some distance, but I could not understand how it could be heard above the continuous and very loud traffic roar of the M25. However in 1995 H.C.Eve told me that he had first heard it calling from the embankments of the M25 at Wisley when he was stuck in a traffic jam. Later he heard it from the same place whilst driving with the windows open at 80 k.p.h. on the M25. Unfortunately the M25 has recently been widened at this point and the embankment has gone.

Then in 1998 I heard from M.Joseph that he had known the Wisley colony for some time and had found the insect to be widespread not only in the village but also around Church Farm, around the sewage farm, along the river Wey and even to the north of the M25 in the Sanway area of Byfleet. It occurred very close to the M25, on both sides, which explained the previous reports from car-drivers. At the very end of the survey period, in October 1998, I visited the area again and heard it calling from the Sanway Road site and also from rough vegetation behind the church adjoining Wisley golf course.

It was now clear that this colony was spread over a large area but it still seemed strange to me that there were no known reports of such an impressive and rare (inland) insect prior to the 1980s. I wondered whether it could have been accidentally introduced into R.H.S.Wisley on plants and had then spread to the village and beyond. In November 1998 I visited the R.H.S. and Andrew Halstead, who told me that it had never been recorded from the Gardens themselves. We then looked through the Fox-Wilson Collection of insects, which is held at R.H.S.Wisley, and to our surprise we found five specimens of Great Green Bush-cricket, all from Surrey. There were two females from Wisley in 1920 and 1922 and three more from Byfleet Sewage Farm in 1928 with a note that it had been abundant there in 1932. At last this was the proof that the Wisley colony was a long-established one and was probably a relic colony from times when this bush-cricket was still common and widespread inland.

There were also two other rather vague reports in about 1980. One was of this bush-cricket having been heard on the roadside at Alfold Crossways. I have searched this locality and only heard Dark Bush-cricket in abundance. The other report was of a nymph from Bowles Wood near Cranleigh, found by a lepidopterist. This locality did look suitable, but I could only find Dark and Speckled Bush-crickets there. Possibly the recorder had mistaken a nymph of the Speckled or Oak Bush-cricket for that of the Great Green Bush-cricket.

The adults are not easy to find, even when stridulating, except at night by torch-light. They stridulate mainly at dusk and at night. The nymphs are rather easier to find, partly because they are more numerous and partly because they tend to be in rough grass rather than in scrub and bushes.

Pholidoptera griseoaptera (De Geer, 1773)   PLATE 3   Dark Bush-cricket

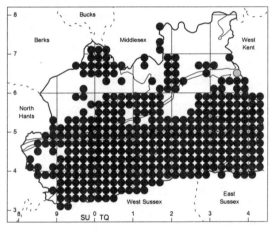

The Dark Bush-cricket is dark brown in the male, pale brown in the female, with a yellow-green underside and virtually no wings. It is quite a bulky-looking medium-sized bush-cricket and is a familiar sight in some gardens.

This bush-cricket is usually regarded as the commonest and most widespread of the bush-crickets, but probably only because it is more easily found than the Oak and Speckled Bush-crickets. Being bulky, conspicuous and not strictly nocturnal, it is often easy to find, and it also has a loud and distinctive, trisyllabic song that it is difficult to miss at night unless, like me and many older people, the songs of all bush-crickets have become inaudible. However the use of a bat-detector will soon show that it is to be found in nearly every bramble-thicket, hedgerow and nettle-bed, and is particularly fond of damp places, often being found by ditches and streams. It is less common in suburban areas, where such areas are often "tidied up" by gardeners or local councils. It normally starts to call at around midday, then sings throughout the latter part of the day and well into the night (all night in high summer if the weather is warm).

Nationally the Dark Bush-cricket is very common throughout southern England and Wales, but scarcely found north of the Wash.

In Surrey it is widespread and common on the escarpment of the North Downs and southwards but its distribution is distinctly patchy north of the Downs. We believe that the map gives a true representation of its distribution. One or two sites may have been overlooked (particularly in the north-west) but, on the other hand, some tetrad dots originate from small isolated colonies covering only a tiny part of the area. As the map shows, there are clearly areas where it is absent. In the west these areas are the acid heaths, where it is too dry and there are few bramble patches or nettle-beds. In the centre there are gaps on the chalk around Effingham and on the downs south of Epsom where it is absent from a wide area, probably because these areas also are too dry; in contrast it is common on the downs around Caterham. It appears to be absent from both Inner and Outer London, although there is still some suitable habitat in allotments and cemeteries and on railway embankments. The nearest locality to central London is probably Ham. It is often abundant on the weald clay, along hedgerows and the edges of woods, and in woodland clearings as well as in gardens. It clearly prefers rather luxuriant vegetation, which is missing from acid heathland and dry upland chalk. It may even be retreating from the drier ground, where patches of thistles and rough grassland are now dominated by Roesel's Bush-cricket.

The continuous chirp of this bush-cricket from overgrown ditches and hedges makes a delightful musical accompaniment to nocturnal walks along country lanes in the Weald or north of the Downs. The song, although loud, is high-pitched and inaudible to many older people; however it is very easily heard on a bat-detector which makes for fast and efficient recording. Almost all the records on the distribution map were made by hearing its song or through the use of bat-detectors. It calls on a frequency of between 20 to 25 kHz, although it can also sometimes be heard on the bat-detector at about 45 kHz. When males meet each other they interact with a rapid churring, which may initially be confused with other species. The first nymphs, which are distinctive and easily recognisable, may be found as early as late April (earliest date 22 April 1990). The adult males can be heard occasionally by mid-July but not normally until the beginning of August (earliest date 14 July 1990); they continue singing until October and sometimes into November.

This bush-cricket appears to be just as common and widespread today as it was last century, and it is possibly even more common now. Burr (1902) said that it was common everywhere, but his notes are so vague as to be almost useless. Lucas (1920), who lived at Kingston-on-Thames, said it "is usually considered to be common but personally I have met with it only in the New Forest area". He then gives a few localities sent to him by Burr, including one "near Wimbledon 1901". It now occurs at Kingston and Surbiton and it seems very strange that Lucas had never seen it in Surrey, unless it was less common in his day, or it had not spread so far north as Kingston.

Undoubtedly the best method of finding this species is by listening for its call (or using a bat-detector) in the evening and at night, when the chirp is most distinctive and can be picked up from 20 metres or so. In the autumn the females are fairly easy to find by walking through long coarse grass near hedges or fields or on the Downs. Another very profitable method is to search brambles and nettles in sunny clearings, ditches etc. in the spring, as the nymphs are easy to see while sunning themselves on the leaves, and are of course much more numerous than later on in the season. They are easy to identify even as small nymphs. Sweeping can also be resorted to, but not in brambles.

Late Dates. A female in a house at Milford on 5 November 1974 (DWB) and a male still calling at Newdigate on 10 November 1984 after ground frosts (RDH).

[*Platycleis albopunctata* (Goeze, 1778)   Grey Bush-cricket

The record for this species, as *grisea occidentalis*, from Cheam in 1944 (Airy Shaw, 1945) was a misidentification of *Metrioptera roeselii* and was later corrected (Menzies and Airy Shaw, 1947). There is no evidence that the Grey Bush-cricket has ever occurred in Surrey.]

Metrioptera brachyptera (Linnaeus, 1761)   PLATE 4   Bog Bush-cricket

Nationally Scarce B

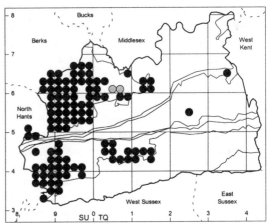

The Bog Bush-cricket is a medium-sized species, either entirely brown or brown with green markings, but always with a green underside. It has a distinctive pale cream margin along the rear edge of the pronotal side-flaps, and only very short wings, except in the extremely rare long-winged form *marginata* (Thunberg). Its song is a continuous chuffing noise, rather like that of a steam train, on a frequency of 25 kHz.

This striking insect is almost restricted in this country to lowland heaths and heathy woods. It is most abundant in the New Forest, east Dorset and west Surrey, but there are a few scattered colonies as far north as Yorkshire and Kirkcudbrightshire.

In Surrey it is most abundant in wet bogs and damp heaths but it does also occur on dry heathland where the heather is tall and dense. It also occurs on the greensand hills up to a height of 294 metres at Leith Hill, 272 metres at Gibbet Hill near Hindhead, 179 metres at Hydons Ball, Hambledon, and 162 metres at Crooksbury Hill. In these localities it occurs in moist upland woods where there is a good undergrowth of bilberry with scattered heather. These upland bilberry woods seem to constitute a unique habitat for this bush-cricket as it has not been reported from such habitat outside Surrey, except on the south side of Black Down in West Sussex at 280 metres. In 1989 Roger Hawkins heard a male calling from a clump of tufted hair-grass by a flooded sandpit at Frimley Green, over a kilometre from the nearest heathland habitat, and returned the following year to confirm its identity by capturing and examining a male.

The very rare macropterous form *marginata* has been recorded twice in Surrey: a female at R.H.S. Wisley Gardens on 3 August 1921 by G.Fox-Wilson, and two males at Thursley Common on 28 August 1970 by M.Chinery & J.Meadows (Ragge, 1973).

The map shows very clearly how closely this species follows the distribution of heathland and especially of cross-leaved heath. The main concentration is on the acid heathlands north and south of the Hog's Back, where it abounds in the wet bogs and damp heaths, with a small outlier at Caesar's Camp near Hale. There is a smaller concentration in upland woods along the greensand ridge from Blackheath, through Winterfold and Hurtwood to Leith Hill and Coldharbour Common, with small outliers at Hydon's Ball and Bourne Woods south of Farnham. It is absent from the south-east of the county but was found in 1983 just outside the county boundary at Copthorne Common (RDH).

Further east there are outlying colonies in damp heathland at Wisley Common, Esher Common, Oxshott Heath and Arbrook Common. In 1983 one was found in an isolated roadside patch of heathy wood at Burwood Park, Weybridge (DWB). This area is now very built-up but there are still remnants of what was once heathland, and the species could be more widespread, especially at St. George's Hill to the south where there is plenty of suitable habitat on and around the golf course, but it is a difficult site to work, being private land. K.G.Blair found it in 1930 at Byfleet, which is on the western edge of St. George's Hill, and J.A.Whellan saw it by the Byfleet Canal in 1945.

East of this there are only two very isolated and vulnerable small colonies. One is on an open heathy area on the east side of Banstead Heath, near Kingswood, where Roger Hawkins found it in 1982. Scrub and trees have invaded this area but the colony was just clinging on in 1998 in the small remaining patch of heather (GAC). This area is in urgent need of proper management. The other is at Addington Hills (sometimes called Shirley Hills), which was first found by Geoffrey Collins in 1943, but this small colony declined in the mid 1950s as the heather became shorter. In 1960 R.A.Farrow reported fewer than six males stridulating on a warm day over the whole locality. The long heather and grass was disappearing rapidly and the crickets were present in birch and gorse bushes. In 1961 the colony appeared to be extinct altogether. However, the open area was retained by management and accidental fires, and the colony persisted. In 1980 Roger Hawkins found eight singing males and one female, all in short heather between paths. In 1991 it was found in long grass rather than short heather, and it was still present in 1998 (GAC). Both these colonies need to be carefully conserved if they are to survive. Once an isolated colony becomes extinct, this species, being flightless, cannot recolonise the site if there is no suitable habitat between the extinct colony and the next nearest colony.

Stephens (1835) gives "near Ripley", which probably refers to Wisley Common. Burr (1902) reports it from many places including Leith Hill. Lucas (1920) gives various localities including Wisley, Esher Common and Leith Hill.

There is no doubt that the Bog Bush-cricket has disappeared from some of its localities through change of habitat, caused mainly by the invasion of heathland by birch, pine and scrub, but on the whole it is as common as ever, mainly because its habitat is unsuitable for building or agriculture, and because it can survive on heathland that is being encroached upon by pine and birch, at least for some years.

Late date. A pair was seen on Witley Common on 13 November 1977 (DWB).

Metrioptera roeselii (Hagenbach, 1822) PLATE 4   Roesel's Bush-cricket

Nationally Scarce B

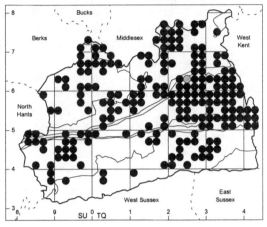

Roesel's Bush-cricket is medium-sized, green and brown with a yellow underside, and with short wings, resembling a long Bog Bush-cricket. However it is distinguished by the cream margin which extends right round the pronotal side-flaps, rather than just at the rear, and by the conspicuous pale yellow spots further back on the sides of the thorax. There is a long-winged form *diluta* (Charpentier), which is quite frequent, especially in hot summers. The song is a very distinctive and very loud, continuous high-pitched buzz sounding like the discharge crackle of overhead electricity lines. It occurs in rough grassland, both coastal and inland.

Until about 1940 this bush-cricket was only known in Britain on both sides of the Thames estuary and up the east coast to the Humber. At about that time it started spreading westwards around London; recently it has spread more rapidly and widely and has been recorded westwards into Oxfordshire in the Thames Valley, north into Huntingdonshire and south into Sussex. It has now also been found in apparently isolated colonies in Hampshire, Somerset, Wales, Lancashire, Yorkshire and even southern Ireland.

It was first found in Surrey at Cheam on 8 September 1944 by the young Ian Menzies who took the specimen to H.K.Airy Shaw. He passed it on to an expert, T.Bainbrigge Fletcher, for identification but, being the then rare long-winged form and also somewhat battered, it was misidentified as *Platycleis grisea occidentalis* (now *albopunctata*) the Grey Bush-cricket (Airy Shaw, 1945). However Ian Menzies found further specimens at the same site in 1946 and correctly identified these as Roesel's Bush-crickets. The original 1944 specimen was then sent to Dr. B.P.Uvarov at the Natural History Museum, London, who confirmed that it was indeed Roesel's Bush-cricket (Menzies & Airy Shaw, 1947). The Cheam site was a rough grass verge separating a field from a minor road but by 1947 this had been set aside for building.

It was also found by others in the next five years at a few more places nearby, but it did not spread further until after 1980 when it moved westwards and then southwards. It is now widespread throughout the county, being common to abundant in suitable habitat in the north, but distinctly less common in the south where it has only recently arrived. During the period of expansion large numbers of the normally rare long-winged form *diluta* have been seen, especially as singletons well away from the nearest colonies. These long-winged forms have been particularly numerous in the hot, dry summers of 1983-4, 1989-91 and 1995-97. It seems that hot summers cause an increase in numbers in the colony which in turn produce many long-winged forms which can fly long distances and start new colonies.

This rapid spread has no doubt been helped by the sudden and recent increase in rough grassland, which is its favoured habitat. The building of the M25 through the middle of Surrey immediately created a linear habitat of ideal south-facing undisturbed rough grass slopes; also many fields which had become cut off by the M25 became abandoned, and these again created ideal habitat. Recently many acres of farms on marginal soil have been turned over to "set-aside" and have reverted to rough grass; there has also been an increasing number of deserted or derelict farms on poor soil. All these have helped Roesel's Bush-cricket (and the Long-winged Cone-head) to spread.

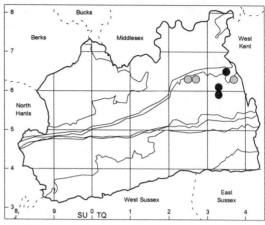

Distribution at the start of the survey in 1970.

Its preferred habitat is rough, rank fields with plenty of tall weeds such as ragwort and thistles, from which the males call; it also occurs in rough grassland both on the chalk and on the London clay and sands. It often occurs on roadside verges and motorway embankments and can obviously tolerate traffic fumes well.

After its discovery at Cheam on a roadside verge in 1944, it was found on a field-margin at Addington in 1948 (GBC), but both sites were destroyed shortly afterwards by building and ploughing. However Ian Menzies rediscovered it at Cheam in 1989, at a site very close to his original discovery; it also now occurs again at Addington. In about 1949 it was also found around Sanderstead (GBC, DRR) and Riddlesdown (RWJU). When Roger Hawkins first visited this site in 1978 the colony was in a very restricted area, but it extended gradually as farming operations ceased on the nearby fields and it is now very common and widespread there. In 1966 it was also recorded at Carshalton, and at about the same time in Lloyd Park near Croydon, where a colony has persisted although reduced by mowing (GBC).

In the 1980s it started to spread slowly westwards around London, becoming widespread at such sites as Wimbledon Common (RDH, 1982), Richmond Park, Barnes Common, Mitcham Common and Ham Common (all MJS, 1984),

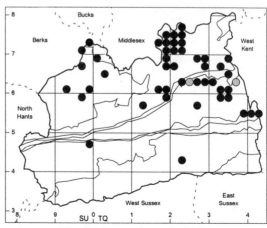

Distribution at the end of 1991.

Runnymede (RDH, 1987), Stroude near Virginia Water (RDH, 1988) and Surbiton (ISM, 1989). There must have been a large expansion in about 1990 because in 1991 it was recorded from almost all places around London with suitable habitat and often in abundance, e.g. at South Norwood Country Park, the former Croydon Airport, Chipstead Valley, Heavers Meadow at Selhurst, Morden Hall Park, Nonsuch Park, Bookham Common and Ashtead Common.

Further south there were records at this time along the M25 embankment near the Kent border (ISM, 1990), although its song was difficult to hear above the deafening roar of traffic. Further west it was also found at Windsor Great Park (MJS, 1991), Chobham Common (MJS, 1991), Horsell Common (DWB, 1990), Knaphill (DWB, 1991), Addlestone, where it was abundant in abandoned fields beside the M25 in 1991 (DWB), and Greyspot Hill near Lightwater (1991).

There were also two isolated records in the south at this time. Roger Hawkins saw and heard a single male f.*diluta* in a damp corner of a field at Stanhill Court near Charlwood in 1984; the species was never seen again at this site and was not discovered in the surrounding area until 1998. Michael Skelton and I saw and heard another single male f.*diluta* at Shalford Meadows near Guildford in August 1991, although it was a few years before this locality was colonised.

Comparing the 1991 distribution map (previous page) with that of the Long-winged Cone-head at the same period (facing page) shows how the cone-head was spreading from the south, whereas this bush-cricket was spreading from the north-east, and their ranges had not quite overlapped.

After 1991 it expanded its range westwards and southwards more rapidly but usually only after a hot summer; sometimes a new colony would die out after a year or two if there was a poor summer. The last big expansion probably took place in 1997 and by the end of the survey it had spread right down to the borders of Sussex and Hampshire; it is still however missing from some areas in the extreme south and south-east. It has not spread as rapidly as the Long-winged Cone-head, although they share a similar habitat and are often found together. It is still much commoner in the north than the south, and many suitable fields in the south have to be scarched thoroughly before a small colony is heard in one small area, but once a pair is present it seems that it can colonise the whole field extremely quickly.

This striking bush-cricket is not at all easily seen in rough grassland; although the males often sing from the tops of grass stems or weeds, they usually drop to the ground at any disturbance. However the song is very loud, piercing and continuous, and it is easily recognised on a bat-detector on a frequency of 20 kHz; it sings only by day, and often stops when the sun goes in. Because the song is so high-pitched, the use of a bat-detector is the best method of locating it, especially for the hard of hearing. Those with good hearing can hear it from a moving car at about 25 metres.

Late Date. One was singing at Sanderstead on 7 October 1991(DWB).

Conocephalus discolor (Thunberg, 1815)   PLATE 5   **Long-winged Cone-head**
[=*fuscus* (Fabricius, 1793)]

Nationally Scarce A

The Long-winged Cone-head is slender and bright green with a brown stripe on top and long brown wings; some specimens have pale brown bodies. This species is best separated from the very similar Short-winged Cone-head by the very long and almost straight ovipositor in the female and by the continuous, high-pitched buzzing song in the male, which is very faint and inaudible to most people. It occurs in coarse vegetation such as rough fields and ungrazed downland.

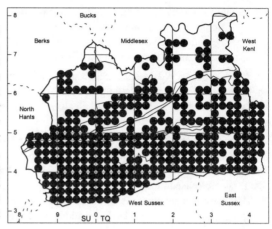

This species was only discovered in England as recently as 1931 on the Isle of Wight; Ian Menzies first found it on the mainland at a coastal site in West Sussex in 1945, while it was found in Dorset in 1947 and in Hampshire in 1970. During the 1980s it started to spread inland, especially through the New Forest.

In Surrey this newcomer was first found in 1990 by three different recorders in three separate localities. The first to find it was Ian Menzies, who was trying out a bat-detector for the first time at Bookham Common on 12 August 1990. He found a male cone-head, which to his surprise flew off, but he assumed that it was a long-winged form of the Short-winged Cone-head. However, on listening to its song with his bat-detector, he realised that it was a Long-winged Cone-head and confirmed this by capturing a specimen.

On 23 August 1990 I was investigating some rough marshy water-meadows by the River Wey at Somerset Bridge, Elstead, searching for the Short-winged Cone-head. I soon found a female which I looked at quite carefully but did not catch; as it had long wings I assumed it was merely a long-winged form, but on the way home I remembered that it had had a very long and rather straight ovipositor. I therefore returned in the afternoon but in spite of an hour's search could find no cone-heads, because at that time I

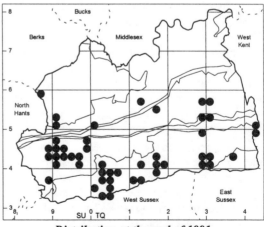

Distribution at the end of 1991.

had no bat-detector. Four days later I was walking round Pudmore Pond on Thursley Common, looking at the tussock sedge for cone-heads. To my surprise I saw and caught a male with long wings which I later identified at home as a Long-winged Cone-head. I immediately went back to Somerset Bridge and after another long search eventually found a large colony of Long-winged Cone-head, thus confirming my original find of four days earlier. The following year I discovered, with the help of a bat-detector, that both cone-heads are abundant in these water-meadows.

On 6 September 1990 Michael Skelton found a few singing in Shalford Meadows, just south of Guildford. The following year he and I found it there again; by then both cone-heads were common and widespread all along the River Wey from Guildford to Godalming and even beyond to Elstead.

Although it was only first recorded in 1990, by the end of 1991 it had been found in numerous sites, mainly in the south-west, but also in the south-east, and one as far north as Farthing Downs at Coulsdon. The first map shows its distribution at the end of 1991, and it is clear that it had colonised much of the southern half of the county within the space of one year. By 1994 the colonisation of the whole county was virtually complete, but numbers continued to increase until in 1997 and 1998 it could be found in almost every patch of long grass that was examined.

Although it occurs with the Short-winged Cone-head in the same habitat of damp rushy fields, it is probably more common in rather drier habitats such as abandoned fields and farms and rough grassland along the North Downs and can even be found in isolated, small patches of long grass or weeds. In fact almost every rough field in Surrey seems to hold large numbers of it. It can apparently produce very large numbers in a colony in a very short time so that a derelict farm can be quickly covered in enormous numbers. A large proportion of specimens have extra-long wings (extra-macropterous), especially in hot summers, and these must help dispersal. The recent series of hot summers has no doubt helped to build up numbers, and the increase in the number of set-aside fields and derelict farms has also helped the phenomenally rapid spread of this species.

It may have spread north from the long-standing Sussex and Hampshire colonies, but there is speculation that this recently fast-expanding population originates from widespread immigration from continental Europe. There are two reasons for this: firstly, that the species has, apparently, arrived in the Scilly Isles, Cornish mainland and Devon from the south, and secondly, that the brown form, which has always been reasonably common in the original English populations, has yet to be recorded in the newly colonised areas. Exactly when it reached Surrey makes for further interesting speculation. Some of the colonies found in 1991 were so large that it must have been there for at least a few years, unless it can colonise a whole farm in one or two years.

As will be seen from the second map, it is now widespread throughout the county, although there are more records from the south than from the north. The insect has spread so rapidly in a few years at the end of the survey period that our recorders have not been able to keep up with it. The apparent absence from parts of the north-west of the county almost certainly reflects lack of recording, rather than lack of cone-heads, and it probably occurs in almost every tetrad outside the built-up areas of Inner London. It is abundant at Richmond Park,

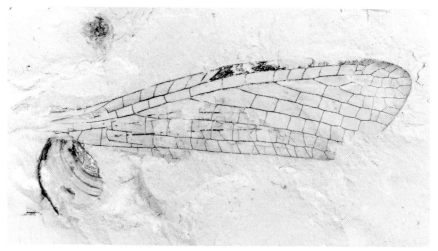

Figure 1. Hindwing of *Panorpidium* sp. (Elcanidae) alongside conchostracan (clam shrimp). Auclaye Brickworks. (From Jarzembowski, 1991.) Length 16mm.

Figure 2. Fore and hindwing of male prophalangopsid. Auclaye Brickworks. BMB 014910 coll. Rita Batchelor. Length 7cm.

Fossil specimens

Figure 3. Stridulatory file on male baissogryllid forewing illustrated in Jarzembowski (1984). Auclaye Brickworks. Scale line in mm.

Figure 4. Locustopseid forewing, Upper Weald Clay. Smokejacks Brickworks. (After Jarzembowski & Coram, 1997.) Length 16 mm.

Fossil specimens

GRASSHOPPERS AND CRICKETS OF SURREY

Meconema thalassinum Oak Bush-cricket (m)

Meconema thalassinum
Oak Bush-cricket (f)

Tettigonia viridissima
Great Green Bush-cricket (m)

Tettigonia viridissima
Great Green Bush-cricket (f)

Pholidoptera griseoaptera
Dark Bush-cricket (m)

Pholidoptera griseoaptera
Dark Bush-cricket (nymph)

Pholidoptera griseoaptera
Dark Bush-cricket (f)

Bush-crickets (Tettigoniids)

PLATE 3

GRASSHOPPERS AND CRICKETS OF SURREY

Metrioptera brachyptera Bog Bush-cricket (m)   *Metrioptera brachyptera* Bog Bush-cricket (f)

Metrioptera brachyptera   *Metrioptera brachyptera*
Bog Bush-cricket   Bog Bush-cricket (nymph)
(nymph)

Metrioptera roeselii   *Metrioptera roeselii*
Roesel's Bush-cricket (m)   Roesel's Bush-cricket (f)

                                                 Metrioptera roeselii
Metrioptera roeselii   Roesel's Bush-cricket (m)   Roesel's Bush-cricket (nymph)
                         (long-winged form)

Bush-crickets (Tettigoniids)

PLATE 4

GRASSHOPPERS AND CRICKETS OF SURREY

Conocephalus discolor Long-winged Cone-head (m)

Conocephalus discolor Long-winged Cone-head (f)

Conocephalus discolor
  Long-winged Cone-head (m)   *Conocephalus discolor* Long-winged Cone-head
  (extra-long-winged form)   (nymph)

Bush-crickets (Tettigoniids)

PLATE 5

GRASSHOPPERS AND CRICKETS OF SURREY

Conocephalus dorsalis Short-winged Cone-head (m)

Conocephalus dorsalis Short-winged Cone-head (f)

Conocephalus dorsalis Short-winged Cone-head (f) (long-winged form)

Conocephalus dorsalis Short-winged Cone-head (nymph)

Conocephalus dorsalis Short-winged Cone-head (mating pair)

Conocephalus dorsalis Short-winged Cone-head (female ovipositing)

Bush-crickets (Tettigoniids)

GRASSHOPPERS AND CRICKETS OF SURREY

Leptophyes punctatissima Speckled Bush-cricket (m)

Leptophyes punctatissima Speckled Bush-cricket (f)

Leptophyes punctatissima
Speckled Bush-cricket (nymph)

Leptophyes punctatissima
Speckled Bush-cricket (male nymph)

Lepto phyes punctatissima Speckled Bush-cricket (m)

Whitethroat *Sylvia communis* with Speckled Bush-cricket

Bush-crickets (Tettigoniids)

PLATE 7

GRASSHOPPERS AND CRICKETS OF SURREY

Acheta domesticus   House Cricket (f)

Nemobius sylvestris   Wood Cricket (f)

Gryllus campestris   Field Cricket (m)

Gryllus campestris   Field Cricket (f)

Gryllus campestris   Field Cricket
(pair at burrow entrance)

True crickets (Gryllids)

PLATE 8

GRASSHOPPERS AND CRICKETS OF SURREY

Gryllotalpa gryllotalpa Mole-cricket   *Gryllotalpa gryllotalpa* Mole-cricket (nymph)

Gryllotalpa gryllotalpa Mole-cricket   Mole-cricket eggs *in situ*

Tetrix subulata Slender Groundhopper   *Tetrix subulata* Slender Groundhopper

Tetrix undulata Common Groundhopper   *Tetrix undulata* Common Groundhopper

Mole-crickets (Gryllotalpids) and Groundhoppers (Tetrigids)

PLATE 9

Stethophyma grossum
Large Marsh Grasshopper (m)

Stethophyma grossum
Large Marsh Grasshopper
(nymph, pink form)

Stenobothrus lineatus
Stripe-winged Grasshopper (m)

Stenobothrus lineatus
Stripe-winged Grasshopper (f)

Omocestus rufipes Woodland Grasshopper (m)

Omocestus rufipes
Woodland Grasshopper (f)

Grasshoppers (Acridids)

GRASSHOPPERS AND CRICKETS OF SURREY

Omocestus viridulus
Common Green Grasshopper

Omocestus viridulus
Common Green Grasshopper (female ovipositing)

Chorthippus brunneus   Field Grasshopper
(purple form)

Chorthippus brunneus
Field Grasshopper (brown form)

Chorthippus brunneus   Field Grasshopper

Chorthippus brunneus
Field Grasshopper (buff form)

Chorthippus brunneus   Field Grasshopper
(straw form)

Chorthippus brunneus   Field Grasshopper
(courtship behaviour)

Grasshoppers (Acridids)

PLATE 11

Chorthippus albomarginatus Lesser Marsh Grasshopper
(green form)

Chorthippus albomarginatus
Lesser Marsh Grasshopper

Chorthippus parallelus Meadow Grasshopper
(courtship behaviour)

Chorthippus parallelus Meadow Grasshopper
(purple form)

Chorthippus parallelus Meadow Grasshopper
(nymphs)

Grasshoppers (Acridids)

GRASSHOPPERS AND CRICKETS OF SURREY

Gomphocerippus rufus Rufous Grasshopper (courtship behaviour)

Gomphocerippus rufus Rufous Grasshopper (nymph)

Gomphocerippus rufus Rufous Grasshopper (m)

Gomphocerippus rufus Rufous Grasshopper (f) (courtship behaviour)

Grasshoppers (Acridids)

GRASSHOPPERS AND CRICKETS OF SURREY

Myrmeleotettix maculatus Mottled Grasshopper (striped form)

Myrmeleotettix maculatus Mottled Grasshopper (f) (orange form)

Myrmeleotettix maculatus
Mottled Grasshopper (f) (green form)

Myrmeleotettix maculatus Mottled Grasshopper (mating pair)

Myrmeleotettix maculatus
Mottled Grasshopper
(two females ovipositing)

Grasshopper exuvium

Grasshoppers (Acridids)

PLATE 14

Ectobius lapponicus Dusky Cockroach (m)   *Ectobius lapponicus* Dusky Cockroach (f)

Ectobius pallidus Tawny Cockroach (m)   *Ectobius pallidus*
                                            Tawny Cockroach (f)

Ectobius panzeri
Lesser Cockroach (female with ootheca)

Ectobius panzeri   Lesser Cockroach (m)   *Ectobius panzeri*
                                             Lesser Cockroach (nymph)

Cockroaches (Dictyopterans)

GRASSHOPPERS AND CRICKETS OF SURREY

Labia minor   Lesser Earwig (m)

Apterygida media   Short-winged Earwig (m)   *Apterygida media*   Short-winged Earwig (f)

Forficula lesnei   Lesne's Earwig (m)   *Forficula lesnei*   Lesne's Earwig (f)

Forficula auricularia
Common Earwig (m)

Forficula auricularia
Common Earwig (f)

Forficula auricularia   Common Earwig
(female with eggs)

Earwigs (Dermapterans)

PLATE 16

Wimbledon Common, Tooting Bec Common and Peckham Rye. It can be difficult to find in parts of the north of the county where it is far out-numbered by Roesel's Bush-cricket, and its calls are swamped by the noise of the larger insect. Although its call is quieter, it can still be traced by tuning the bat-detector to the correct frequency.

It has clearly spread and increased at a phenomenal rate since its arrival in Surrey in about 1990, and is now among the most widespread and common of all the bush-crickets in the county.

Last-instar nymphs can be confusing, since the females are almost as large as the adults and have well-developed ovipositors and short wings, and so may be mistaken for adult Short-winged Cone-heads. On close examination, the ovipositor is found to be long and straight, while the wings are merely short pads.

It is very difficult to see in rough grass and, because it has a high-pitched song which is inaudible to many people, the best way to find this cone-head is by using a bat-detector. Its song is a distinctive continuous call on a frequency of 30 kHz but the note changes briefly as the song slows down just before stopping. It will start singing before midday if the sun is out and will continue late into the evening, especially when warm. It will sing in cool conditions but the song is then much fainter and sometimes slower and it can then sound rather like the Short-winged Cone-head. The males normally call from long grass or thistles, resting along the stems, facing either way up; they hide behind the stem when approached, moving round it as the observer moves. They continue to call when upside down on a stem or even when crawling around. The call of the extra-long-winged (extra-macropterous) form, in which the male's wings extend far beyond the tip of the abdomen and the female's to the end of the ovipositor, is on exactly the same frequency of 30 kHz as the normal form.

It is easily located by merely walking into any suitable rough field and listening for the song with a bat-detector, even though it may be difficult to see.

Late Date. It was still singing at 5pm on 13 October 1991 at Dunsfold in a derelict meadow (DWB).

Conocephalus dorsalis (Latreille, 1804) PLATE 6 **Short-winged Cone-head**

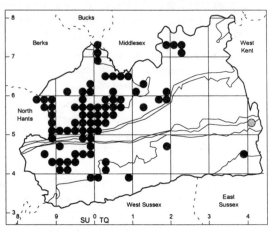

The Short-winged Cone-head is a small, slender, bright green species with a brown stripe on top and short brown wings; there is however a not infrequent long-winged form called *burri* Ebner. Even the normal form is very similar to the Long-winged Cone-head but it is smaller and has shorter wings; however the adult female is best distinguished by its shorter and more upcurved ovipositor and the male by its distinctively different song. The song, which is continuous, high-pitched and so soft as to be almost inaudible, consists of alternating sounds: one a hissing noise like that of the Long-winged Bush-cricket and the other a ticking noise like a fishing reel. It is normally found in damper places than the Long-winged Cone-head, such as salt marshes and estuarine meadows and inland in wet meadows and ditches. Until recently it was mainly a coastal species in southern England and Wales, but it has now expanded its range inland and is widespread in eastern England especially around the Thames Valley.

Before this survey there was only one recent Surrey record (from Limpsfield Common in 1968), but it has now been found to be widespread and locally common in the west. It is assumed that it has spread recently, due to many hot summers, but it is also very likely that it was always present in small numbers locally and was overlooked. It is a small retiring species which moves rapidly through the rushes when disturbed and can easily be missed unless a special search is made for it.

It occurs mainly in damp rushy fields, particularly water-meadows in the Wey valley, but also on damp clay commons and at the edges of ponds and ditches, where there are rushes or tall grasses. It has not yet been found on heathland bogs although it has colonised this habitat in the New Forest, co-existing with the Bog Bush-cricket. In Sussex, Haes (1976) found that it always occurred with the Lesser Marsh Grasshopper, as they both occupy the same habitat. They both occur together in Surrey at various sites such as water meadows in the Wey valley between Pyrford and Send, in the Blackwater valley at Ash Vale and at Greyspot Hill near Lightwater, but the Short-winged Cone-head is absent from many sites in east Surrey where the Lesser Marsh Grasshopper occurs. Haes also found that in Sussex it occurred in reed-beds, but it has not so far been found in this habitat in Surrey.

The long-winged or macropterous form *burri*, with both pairs of wings extending well beyond the abdomen, has been found frequently in recent years. This looks very like the Long-winged Cone-head, especially in the male, which should be carefully checked. The best method of distinguishing them is by listening to the songs on a bat-detector. The songs are usually distinctively different, the present species having a continuous song with the speed changing at intervals. The Long-winged Cone-head does not normally change the speed of the song, but occasionally a few specimens do and then the sound is very similar to that of the Short-winged Cone-head. However the use of a bat-detector solves this problem because the Long-winged Cone-head calls on a frequency of 30 kHz, whereas the Short-winged Cone-head calls on a frequency of 40 kHz, even in its long-winged form *burri*.

It tends to sing only when the sun is out or it is very warm; as soon as clouds obscure the sun, or by late afternoon when it gets cooler, this species stops singing. The Long-winged Cone-head will sing on cloudy days and continue even after dark. Both continue to sing when disturbed or approached very closely, and even when they are moving up or down a rush stem. In the late afternoons I have often noticed both sexes climb up wooden posts and fences, presumably to be able to bask in the sun as long as possible; I once saw as many as six on one fence post.

Although the first modern record for Surrey was from Limpsfield Common (1968), it has not been seen there again, and it appears to be extremely scarce in south-east Surrey, in spite of plenty of suitable habitat there. The only records from the south-east are from Holmwood Common, Arden Green near Lingfield (DWB, 1992) and a singleton male form *burri* at Ewhurst Green (DWB, 1991).

Its main stronghold is the valley of the River Wey, where it is common to abundant in water-meadows from Weybridge to Farnham. It also occurs on most of the damp, rushy, clay commons north and west of Guildford such as Clasford Common near Normandy, Horsell Common, Sow Moor near Chobham, Stringers Common and Backside Common near Guildford, and Fox Corner near Worplesdon. It occurs south of Guildford along the Cranleigh Water, a small tributary of the Wey, and also in an apparently isolated colony at Hascombe.

On the Hampshire border it occurs in the Blackwater valley, from Hale to Camberley, at the edges of ponds and gravel-pits, e.g. Mytchett Lake (RMF, 1980). In the north it occurs at Runnymede (RDH, 1987) and Stroude near Virginia Water (RDH, 1988) and also as near to London as Richmond Park, where it was discovered abundantly in rushes on both sides of Pen Ponds (ISM, 1990), and on Wimbledon Common by two ponds in 1991 (DWB). It was found on Epsom Common in 1984 (MJS), Bookham Common in 1989 (ISM) and Esher Common in 1991 (DT).

Lucas (1920) only gives a single record – "near Witley"; although this is the first published record for Surrey, Lucas does not comment on it. It has been found recently in a rushy field in the parish of Witley near Thursley Common (DWB, 1990). The next record was not until 1968 at Limpsfield Common where it has not been refound. It was next found in a damp, rushy field just north of Guildford in July 1975, where it occurred in abundance over a large area (DWB). In the following two years more colonies were found near Woking, Guildford, Rushett Common near Cranleigh and on the Wey at Godalming. By 1980 it had been found at Mytchett Lake and by 1990 it was being found in almost every rushy field along the Wey valley and in the Blackwater valley. Ian Menzies found it in Richmond Park in the same year.

In August 1991 I borrowed a bat-detector and by using this I found that it was very common in all suitable habitat in west Surrey as well as on Wimbledon Common. I found it in many places where I had failed to find it by sight in previous years. Roger Hawkins, also using a bat-detector, failed to find it in rushy fields in south-east Surrey, so that it does seem to be extremely scarce there; I eventually found a colony near Lingfield with the aid of a bat-detector.

Because it is so difficult to locate by sight and also because it has a rather quiet high-pitched song, it is quite possible that it was overlooked by earlier recorders; its habitat is also one which orthopterists might easily neglect. It may therefore have always been present but in small numbers, although it has clearly increased considerably in recent years and has expanded its range, possibly spreading from the Thames Valley. The series of hot summers and the large numbers of the long-winged form have obviously helped such a spread. The use of a bat-detector shows how the distribution can be recorded in a very short time.

As has been shown above, by far the best way of finding this cone-head is by the use of a bat-detector on sunny days in damp, rushy fields. The sound pattern, on 40 kHz, is very distinctive, and, although soft, it can be picked up from about 25 metres. Even when located by sound, they can be remarkably difficult to find by sight, as they will sometimes sing from well down in the middle of a clump of rushes. Sweeping is also recommended but I have not personally used this method.

Late Date. It was still abundant at Stoke Meadows near Guildford on 25 October 1977 (DWB).

Leptophyes punctatissima (Bosc, 1792)   PLATE 7   Speckled Bush-cricket

The Speckled Bush-cricket is a short-bodied, plump little bright-green bush-cricket with long spindly legs; as its name implies, it is covered with minute dark spots. It has only vestigial wings but the female has a stout, upcurved ovipositor. Its call is a distinctive, very high-pitched tick, repeated at longish intervals; the sound is inaudible to most humans but is very easily picked up on a bat-detector. It is a familiar sight in gardens, bramble patches, nettle beds and other overgrown spots, sunning itself on leaves or flowers, and it occasionally comes into rooms at night. It is very widespread and common in southern England and Wales.

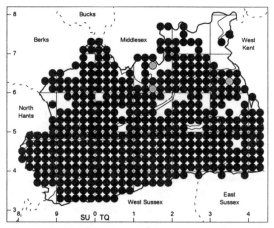

In Surrey this bush-cricket is also widespread and common, and is probably the most frequently seen of the bush-crickets, even though the newly arrived Long-winged Cone-head may now be equally common. It inhabits the same rough bushy vegetation as the Dark Bush-cricket, but it is often found quite high in trees and bushes. Roger Hawkins has located, by means of a bat-detector, males calling from up to 6 metres above ground in ash and lime trees; he has also found them occasionally in conifers, e.g. Lawson's cypress and Scots pine. It is frequently found in gardens, its favourite site being bramble and nettle patches on the south side of a hedge or of bushes, but it is unusual to find more than a few adult specimens at any one site. They may often be seen sitting exposed on a leaf of bramble or hedge bindweed, absorbing the warmth of the morning sun, but as the sun rises higher they disappear beneath the leaves and are rarely seen after midday on fine days. They usually sit absolutely still, except for gentle movements of the antennae, but one male, which was sitting on a creeping thistle near two Dark Bush-crickets, was seen to move in an extremely stealthy manner, consisting of bringing forward each leg in turn in a series of tiny jerks; this slow movement would presumably prevent capture by predators that only see moving objects (RDH). Not much is known about their feeding habits, except that they appear to eat leaves of herbs, e.g. nettles, as nymphs, and of trees, e.g. birch and oak, as adults (Brown, 1983). I have often seen them, both as large nymphs and as adults, eating the petals of rosebuds in the garden.

The nymphs usually hatch in May (earliest date 2 May 1993) and can be found frequently during June and July on brambles and nettles, and at the flowery margins of fields. The first adults may appear in July but most become mature during August. They can still be found in the same situations as the nymphs, but many males climb higher and are often heard or seen at the top of hedges, up to 4 metres above ground. Adults are very common in September and generally live on throughout October, and occasionally into November.

Roger Hawkins has observed females by torch-light on several occasions climbing up tree trunks, or even lamp posts. On 27 September 1984, near Shere, he saw six females, and one male, on various species of tree from 1.5 to 3 metres above ground. Two females were laying in crevices in the bark above forks of the trunk of a horse-chestnut and a young lime; both were about 2 metres above ground, with head down, back arched and the body curved right round. On 30 September 1985, south of Ockley, a female was laying on a trunk of oak, 1 metre up, with its ovipositor inserted into a patch of stunted *Hypogymnia* lichen. The insect faced up the tree with its abdomen doubled underneath it, and made slow up-and-down movements with its body; 20 minutes later it had stopped laying and climbed to 3.5 metres above ground. The next day another was seen inserting its ovipositor into a crack in an ash tree 2.5 metres up. Laying on tree trunks is rarely observed, compared with the Oak Bush-cricket, and it is likely that most of the eggs are laid in bushes or in hedges.

It is a difficult species to record because, in spite of its size, it manages to hide unobtrusively under leaves or flowers. The very high-pitched call is virtually inaudible, but it can be picked up very clearly from up to 30 metres by a bat-detector, the intermittent tick being very distinctive. It calls constantly from dusk until well into the night, but often also by day, rather unpredictably. Sometimes it calls for an hour or two around midday, if it is warm; on other occasions it is active in warm, muggy conditions, and later in the year, when the nights are cold, it may call throughout most of the day. In 1983 Roger Hawkins found that some of his friends were able to hear the calls, although they were totally inaudible to him, so in 1984 he purchased the bat-detector which had been developed by Queen Mary College and put on the market at an affordable price. In the following year he recorded Speckled Bush-cricket on many occasions and was probably the first person in the country to use this device for the systematic field recording of Orthoptera. Males are not easy to locate, even when calling, but on occasions they have been observed moving about on hedges, with antennae active, calling as they moved (RDH). The careful observation of captive specimens has now shown that the male walks towards a female responding to its call (Hartley & Robinson, 1976).

The map clearly shows that it is widespread and common throughout Surrey, occurring in Inner London and in dry heathy country, where the Dark Bush-cricket is apparently absent. It may well occur in every tetrad as it has not been possible to visit all tetrads with a bat-detector. I only started using a bat-detector in September 1991 and in one month I nearly doubled the number of tetrad records that I had made in the previous twenty years. By driving around lanes after dark I heard the distinctive tick from virtually every place that I stopped, and it was quite possible to record it in 25 different tetrads in little more than one hour. The great majority of the records on the maps were made by using bat-detectors.

The localities nearest to central London are Sydenham Hill railway station, Nunhead Cemetery in Peckham, Barnes Common and Wandsworth Common. However it is very likely to occur in the parks and commons, and on railway banks, even nearer to the centre.

All the old authors describe it as common and widespread, giving numerous localities. Its status has probably not changed since it can continue to thrive in gardens, even in built-up areas, and it is common in hedges that are still a familiar feature of the Surrey countryside.

The small distinctive nymphs are easily seen sunning themselves on brambles and nettles in May and June; sweeping rough herbage can also be successful. But by far the best method of finding adults is by using a bat-detector in the evening or at night in late summer and autumn. The very high-pitched tick can clearly be heard from up to 30 metres on 40 kHz and it is repeated about every four seconds, the female responding with a rather softer tick.

Late Date. A female in the garden at Milford on 19 November 1978 (DWB).

GRYLLIDAE – Crickets

Crickets are brown or black compact insects with long antennae and large heads. The males have a very distinctive stridulatory song which they produce by rubbing their forewings together. They tend to be nocturnal or at least crepuscular. Apart from the introduced House Cricket, there are only three crickets native to Britain, two of which are exceedingly rare, and only one now occurs in Surrey, another having become extinct in 1964.

Acheta domesticus (Linnaeus, 1758)   PLATE 8   House Cricket

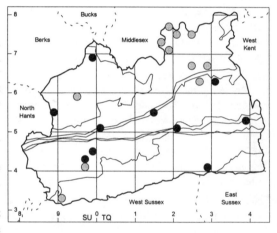

The House Cricket is not strictly a native of Britain but it has been here for so long that it is normally treated as one. It may have been introduced as long ago as the 13th century by the Crusaders from the Middle East from where it is thought to have originated.

It is a medium-sized, greyish-brown cricket, fully-winged in both sexes, which needs constant warmth to be able to survive in this country. It was the familiar "Cricket on the Hearth". Gilbert White of Selborne knew it well and his notes, published in 1789, are worth quoting at length.

"...the house-cricket resides altogether within our dwellings, intruding itself upon our notice whether we will or no. ...it is particularly fond of kitchens and bakers' ovens, on account of their perpetual warmth. These insects, residing as it were in a torrid zone, are always alert and merry: a good Christmas fire is to them like the heats of the dog-days. Though they are frequently heard by day, yet is their natural time of motion only in the night. As soon as it grows dusk, the chirping increases, and they come running forth, and are from the size of a flea to that of their full stature.

...In the summer we have observed them to fly, when it became dusk, out of the windows, and over the neighbouring roofs. This feat of activity accounts for the sudden manner in which they leave their haunts, as it does for the method by which they come to houses where they were not known before. It is remarkable, that many sorts of insects seem never to use their wings but when they have a mind to shift their quarters and settle new colonies. When in the air they move "*volatu undoso*", in waves or curves, like wood-peckers, opening and shutting their wings at every stroke, and so are always rising or sinking.

When they increase to a great degree, they become noisome pests, flying into the candles, and dashing into people's faces; but they may be blasted and destroyed by gunpowder discharged into their crevices and crannies. Their shrilling noise is occasioned by a brisk attrition of their wings. Cats catch hearth-crickets, and, playing with them as they do with mice, devour them."

Although it was once very common and widespread in houses, bakeries and hospitals throughout Britain, it has decreased very rapidly this century and is now probably rare in Surrey, with few definite records since 1980. It does occur outdoors, especially on rubbish-tips, but even here it has not been recorded for some time as rubbish-tips are now much more hygienic and not suitable for crickets.

Burr (1902) described it as abundant in warm places such as kitchens, bakehouses and restaurants, but gave no precise localities. Lucas (1920) noted that it was undoubtedly on the decrease and gave as localities Kingston-on-Thames, 1898; Bisley, 1899; and Haslemere, in a bakehouse, 1908. In 1931 Lucas gave an amusing description of a cricket "making merry for a long time during the evening at a dinner in a Kingston restaurant. The room was large but the sound well filled it although there were many present at dinner".

Between 1940 and 1960 there are many reports from south London by the Ministry of Health, and specimens in the Natural History Museum, London, mainly from hospitals, e.g. Roehampton, Wandsworth, Mitcham, Morden, Kew Gardens, Ham, Queen Mary Hospital at Carshalton, and King George V Sanatorium at Milford where it occurred till 1970 at least. There was a colony at Juniper Hall Field Centre in 1961. Two adults were seen in the Manor Hotel, Farncombe, in April 1973 (DWB) and Judith Marshall saw it at Ash Vale in 1975. Mrs. Morgan reported it from Walton-on-Thames in 1978, one was heard singing in a garden at Fetcham in 1975, and the most recent record was from Oxted Post Office in 1997.

The only rubbish-tip records are from Guildford, Egham and Godalming. There was a very strong colony at the Guildford tip at Stoke Meadows from 1971 until 1976 when the tip was finally closed (DWB). It was also reported from the Callow Hill tip at Egham in 1975 by M.Davies and from the Godalming tip in Ashtead Lane where it was in vast numbers and had spread into the surrounding woods in 1975. Shortly after this the tip was closed and covered over. There was however a colony in 1994 at Betchworth Quarry, when this was being filled (DAC).

There was a most interesting record from Roger Hawkins who reported a strong colony living outdoors in the hot summer of 1983, in the cracks of paving stones at Meadowcroft Close, Horley, where he lives. There were up to 10 males calling from 28 July to 8 September. There are other records in England of colonies living outdoors during the summer, sometimes a long way from buildings, but this cricket is now bred in large numbers for feeding captive reptiles and it is thought that some of these records may refer to escaped breeding stock, as is certainly the case in several counties.

This cricket is almost certainly still surviving in Surrey but definite records would be welcomed. Restaurants, hospitals and rubbish tips are probably the most likely places to look or listen for them. Try also near pet suppliers dealing in reptiles and spiders.

Gryllus campestris Linnaeus, 1758   PLATE 8   Field Cricket

**National Status:
RDB1 (Endangered)**

The Field Cricket is large and bulky, with a big head, and is shiny black with yellow patches at the wing-bases and orange on the underside of the hind-legs. The male has a loud shrill penetrating and continuous song, made by rubbing its large forewings together. It used to occur at the edges of sandy heaths and commons as well as on sunny chalkland slopes in south-east England. It was probably never common, with only a few colonies in Surrey, Sussex, Hampshire and the Isle of Wight. It has decreased rapidly this century to the point of near-extinction with only one colony left in West Sussex.

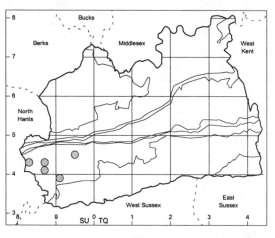

Gilbert White, in his *Natural History of Selborne* of 1789, gives a wonderful description of a colony at Selborne, Hampshire, which is only about 10 kms from the Surrey border, and as this is still possibly the best account of the habits of the Field Cricket it is here quoted almost in full.

"There is a steep abrupt pasture field interspersed with furze close to the back of this village, well known by the name of the Short Lithe, consisting of a rocky dry soil, and inclining to the afternoon sun. This spot abounds with the *gryllus campestris*, or field-cricket; which, though frequent in these parts, is by no means a common insect in many other counties.

As their cheerful summer cry cannot but draw the attention of a naturalist, I have often gone down to examine the economy of these *grylli*, and study their mode of life; but they are so shy and cautious that it is no easy matter to get a sight of them; for, feeling a person's footsteps as he advances, they stop short in the midst of their song, and retire backward nimbly into their burrows, where they lurk till all suspicion of danger is over.

...we learned to distinguish the male from the female; the former of which is shining black, with a golden stripe across his shoulders; the latter is more dusky, more capacious about the abdomen, and carries a long sword-shaped weapon at her tail, which probably is the instrument with which she deposits her eggs in crannies and safe receptacles.

...a pliant stalk of grass, gently insinuated into the caverns, will probe their windings to the bottom, and quickly bring out the inhabitant. ...The males only make that shrilling noise perhaps out of rivalry and emulation. It is raised by a brisk friction of one wing against the other. They are solitary beings, living singly male or female, each as it may happen; but there must be a time when the sexes have some intercourse, and then the wings may be useful perhaps during the hours of night. When the males meet they will fight fiercely...With their strong jaws, toothed like the shears of a lobster's claws, they perforate and round their curious regular cells, having no fore-claws to dig, like the mole-cricket. Of such herbs as grow before the mouths of their burrows they eat indiscriminately, and on a little platform which they make just by, they drop their dung; and never, in the day time, seem to stir more than two or three inches from home. Sitting in the entrance of their caverns they chirp all night as well as day from the middle of the month of May to the middle of July; and in hot weather, when they are most vigorous, they make the hills echo, and in the stiller hours of darkness may be heard to a considerable distance. In the beginning of the season their notes are more faint and inward; but become louder as the summer advances, and so die away again by degrees...the shrilling of the field-cricket, though sharp and stridulous, yet marvellously delights some hearers, filling their minds with a train of summer ideas of everything that is rural, verdurous, and joyous.

About the tenth of March the crickets appear at the mouths of their cells, which they then open and bore, and shape very elegantly. All that ever I have seen at that season were in their pupa state, and had only the rudiments of wings, lying under a skin or coat, which must be cast before the insect can arrive at its perfect state (we have observed that they cast these skins in April, which are then seen lying at the mouths of their holes); from whence I should suppose that the old ones of last year do not always survive the winter. In August their holes begin to be obliterated, and the insects are seen no more till spring.

Not many summers ago I endeavoured to transplant a colony to the terrace in my garden, by boring deep holes in the sloping turf. The new inhabitants stayed some time, and fed and sung; but wandered away by degrees, and were heard at a further distance every morning, so that it appears that on this emergency they made use of their wings in attempting to return to the spot from which they were taken."

Gilbert White also makes the first mention of the Field Cricket in Surrey as long ago as 1761, when in *The Naturalist's Calendar* for that year he writes of it "abounding most in sand-banks on the sides of heaths, especially in Surrey and Sussex." He may well have seen or heard it himself around Frensham which is only about 10 kms from Selborne. Because of White's famous description of his Selborne colony many people have assumed that the Field Cricket was widespread and common in the eighteenth century, but White wrote in his *Naturalist's Journal* for 29 May 1791: "The race of field-crickets, which burrowed in the Short Lythe, and used to make such an agreeable shrilling noise the summer long, seems to be extinct." He also pointed out that, although it was frequent at Selborne, it was not a common insect in other counties.

Unfortunately the Field Cricket is almost certainly now extinct in Surrey.

Burr (1902) in the *Victoria County History of Surrey* says that it is unknown in the county but it should be sought for and "it may be found in warm sandy commons". However in the *Victoria County History of Hampshire* he says it is found at Farnham, which is of course in Surrey, although near the Hampshire border; this locality could possibly refer to the later well-known colony at Frensham, just south of Farnham.

In the K.G.Blair Collection at the County Museum in Winchester there is a specimen of a male Field Cricket labelled "Thursley 24/5/1914" (next to an unlabelled female which may have also come from Thursley). Lucas (1920) did not know of any colony at Thursley or Farnham although he does quote Burr. The only colony which he reported was "between

Eashing and Godalming, whence J.G.Dalgliesh received specimens". I have searched this area and, although there are some steep south-facing slopes with rabbit-grazed turf on Hythe sand, I have failed to find the cricket there. There are six specimens in the Charterhouse Museum Collection but none of them has a data label. They are in a drawer containing other insects collected by O.H.Latter around 1915 in the Godalming area; he was a master at Charterhouse at that time and for many years after. It is tempting to suppose that these specimens came from this colony between Godalming and Eashing, as Charterhouse is midway between the two, but we shall never know unless we find Latter's diary or notes, if indeed he made any. Lucas (1920) also mentions a singleton from Rotherhithe in 1904, but this was presumably an accidental introduction with imported food in the wharves or even a misidentification for an alien imported species.

However a colony was known in the 1950s at Frensham Little Pond and another nearby at Tilford Common; as they were only about 1 km apart they probably formed part of one large scattered colony. Unfortunately the Frensham colony, which was on the east side of the Little Pond and which was thriving in 1952 (many specimens caught by J.Abraham in 1952 and by S.M.Shephard in 1954; also noted by C.Diver), became rapidly extinct, probably due to the semi-commercialisation of the pond and its increasing popularity as a week-end resort (Ragge, 1956). It was finally obliterated by an unofficial car-park in 1954.

The Tilford colony, which was on privately owned heathland and an abandoned building plot beside the road just south of the village, existed until 1964 when Chris Haes and Peter Le Brocq heard two males chirping on 30 May, but it has not been seen or heard there or anywhere else since that date. This colony had been known by Peter le Brocq since 1937, when as a boy he cycled along the road each day to school and back. Being curious as to what was making the shrill noise, he tried pouring water down the burrows, but as this was not successful he then poked stems of wavy hair-grass down them and this method produced the male crickets. It was a strong colony until about 1960, when the owner was unable to maintain the mowing necessary to prevent the encroachment of scrub and bracken onto a formerly rabbit-grazed area; by 1964 the site had become unsuitable because of this encroachment. However, there is still suitable habitat near both these sites.

In 1979 I decided to try to reintroduce the Field Cricket to Surrey because it had probably only recently died out there and was on the way to extinction in Sussex. I picked a south-facing sandy slope on Witley Common covered with heather, interspersed with grass, and conveniently near my home. I obtained the consent of the Nature Conservancy, and also of the National Trust who are the owners of Witley Common.

Chris Haes sent me about 120 third-instar nymphs, bred from a Sussex colony by Mike Edwards, and I released these at three sites on Witley Common on 30 July 1979 and a further 12 ninth-instar nymphs on 20 October 1979. On 10 May 1980 I heard one male singing from one of the release sites. On 26 and 28 May I heard a male, presumably the same one, singing about 100 metres to the south and on 12 June I heard it singing about 200 metres to the west. Unfortunately I never heard it again and presumed it to be an unmated wandering male. It was very reminiscent of the Field Crickets which Gilbert White had tried unsuccessfully to introduce into his garden in 1761, over 200 years earlier; both he and I had "hoped they would increase on account of their pleasing summer song".

I was also sent 15 ninth-instar nymphs which I put into a tank filled with earth on 10 October 1979. Nine survived the winter and the first male moulted into an adult on 20 May 1980. The males started singing on 3 June and sang continuously thereafter. The first female moulted to adult on 12 June, and the last male moulted to adult on 21 June.

As the Field Cricket was on the verge of extinction in Britain by 1990, English Nature chose it for its Species Recovery Programme, which it launched in 1991. Mike Edwards was in charge of the Field Cricket project and he made a very thorough study of its habits and life history, both at its last remaining British colony in West Sussex and at other sites in Europe. The results of this study were published in *British Wildlife* in 1996 under the title "The Field Cricket – Preventing Extinction" (Edwards *et al*, 1996); this paper is the only comprehensive account to have been published since White.

Surveys were made of possible sites for reintroduction or even for new introduction near an historical site, if the original site was now unsuitable. The cricket was introduced to sites in Sussex, Hampshire and Surrey, some successfully and some not. The site chosen in Surrey was the south-facing heathery slope above Frensham Great Pond, which is close to the last-known sites at Tilford and Frensham Little Pond, both of which were by then unsuitable. In the late summer of 1994 Mike Edwards released at the Frensham site about 2000 nymphs, which had been bred from native stock by London Zoo, and in the following May some males were heard or seen. In the subsequent spring one burrow, but no occupant, was found so that breeding must have occurred. In 1997 a further release of 200 nymphs was made to the south, and also near the car-park and to the east of the road; in May 1998 nine males were found. A further 2,000 nymphs were released there in the summer of 1998. He is looking at other possible sites in south-west Surrey for further introductions.

Nemobius sylvestris (Bosc, 1792)   PLATE 8   Wood Cricket

Nationally Scarce A

The Wood Cricket is a small, dark brown cricket, with paler markings, and with very short forewings and no hindwings, which jumps around in the leaf litter in sunny woodland clearings when disturbed. The male has a quiet purring song which is easy to hear but difficult to locate.

It was first recorded in Britain in 1820 from a sandpit at Lyndhurst in the New Forest. It is still in the New Forest, where it is now abundant, but is also common on the Isle of Wight and in parts of South Devon. There is a suspicion that, although it may be native on the Isle of Wight, it may have been accidentally introduced elsewhere on trees from the continent.

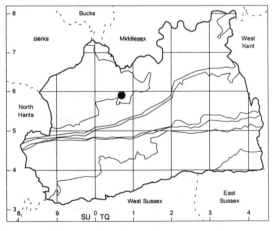

The same suspicion attaches to the only colony of Wood Crickets in Surrey at Wisley. This was first discovered in 1967 by A.E.Stubbs who came across two small colonies on Wisley Common in the woods around the car-park of the R.H.S. Gardens (Stubbs, 1967). The habitat was very similar to that of the New Forest colonies but this locality was about 80 kms from the nearest one in the New Forest. Wood Crickets cannot fly and it was therefore assumed that specimens had been introduced accidentally on plants from the New Forest, possibly with some azaleas from Exbury in 1963, when several crickets were seen by Chris Haes on a consignment being assembled for Wisley Gardens; if so the colony had quickly spread. However at that time it was not known from within the Gardens themselves, although it now occurs in the woods between the Alpine Garden and the river. It is also possible that the colony had remained undetected on the common until 1967 and that intervening colonies in north-east Hampshire and south-west Surrey have also not been detected or have become extinct – there is still plenty of suitable habitat in those areas and the insect is very elusive unless its very soft purring call is heard. Azaleas and other shrubs and trees have been exported for years from Exbury and it seems odd that the Wood Cricket has never been found at other suitable sites in the country.

When Stubbs published his discovery in 1967 in *The Entomologist*, he did not disclose the actual site, but in August 1976 I happened to be searching for grasshoppers on Wisley Common when I heard to my great surprise the faint but unmistakable song of Wood Crickets. They were plentiful over an area of about 200 square metres in open woodland adjoining the R.H.S. car park and also on the edge of Wisley Common, 500 metres to the north, in presumably the same site as the original discovery. I saw them there for a year or two afterwards, but the car park was then extended and much of the site was destroyed, and I feared that the colony had gone. However it has been recorded in recent years, by various people, as occurring plentifully in the woods on the north side of Wisley Common known as Ockham Forest and over a wide area towards Boldermere, so that it appears to be in no danger and is possibly spreading. However the site is now bounded on the north by the M25, on the south and east by the A3 and on the west by open country. These two roads could prove to be impassable to the flightless Wood Cricket but to my surprise I recently heard from David Boddy, the Ranger for Wisley Common, that he has recorded it in recent years from the woods just to the north of the M25 which are part of Wisley Common. He thinks that it may also have been seen or heard in the woods on the south of the A3 opposite the road to Wisley and it could be well worth looking for on the adjoining Ockham Common, although this common is mainly dominated by pine trees with very few oaks. He also observed that the Wood Cricket was particularly abundant in 1998 on Wisley Common and even occurred on the open heathland.

The Wisley colony is therefore clearly well-established and extensive and may even be spreading, but we shall never know for certain whether it is native or introduced.

Listening for the song at the edge of sunny clearings or by rides in deciduous woods is the best method of finding this cricket, but failing that, kicking through deep leaf litter in similar places should reveal the insects, which hop actively when disturbed. Nymphs were found quite commonly in leaf litter around birch trees on Wisley Common in February 1999 by Graham Collins.

GRYLLOTALPIDAE – Mole-crickets

The Mole-crickets are very large and distinguished from all other Orthoptera by their highly modified forelegs. They have short antennae and a large pronotum. They are mainly carnivorous, eating worms and caterpillars, but will also eat plant roots. They have a very distinctive continuous resonant call which the males make from a specially contructed resonating calling chamber just below the surface of the soil, which causes the noise to be loud and far-reaching. There is only one species in Britain but at least one other in Europe.

Gryllotalpa gryllotalpa (Linnaeus, 1758)   PLATE 9   Mole-cricket

National Status: RDB1 (Endangered)

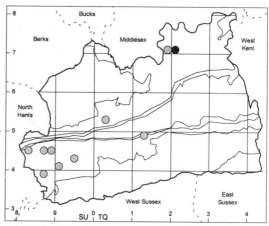

The Mole-cricket is an unmistakable large, pale chestnut-coloured insect, with large mole-like front legs, looking unlike any other cricket. Once again the best description of this insect and its habits is by Gilbert White in his *Natural History of Selborne*, published in 1789; he was obviously familiar with it as there are many references to it in his *Naturalist's Journal* going back to 1771. He writes:

"...while the field-cricket delights in sunny dry banks, and the house-cricket rejoices amidst the glowing heat of the kitchen hearth or oven, the *gryllus gryllotalpa* (the mole-cricket) haunts moist meadows, and frequents the sides of ponds and banks of streams, performing all its functions in a swampy wet soil. With a pair of fore-feet, curiously adapted to the purpose, it burrows and works under ground like the mole, raising a ridge as it proceeds, but seldom throwing up hillocks.

As mole-crickets often infest gardens by the sides of canals, they are unwelcome guests to the gardener, raising up ridges in their subterranean progress, and rendering the walks unsightly. If they take to the kitchen quarters, they occasion great damage among the plants and roots, by destroying whole beds of cabbages, young legumes, and flowers. When dug out they seem very slow and helpless, and make no use of their wings by day; but at night they come abroad, and make long excursions, as I have been convinced by finding stragglers, in a morning, in improbable places. In fine weather, about the middle of April, and just at the close of day, they begin to solace themselves with a low, dull, jarring note, continued for a long time without interruption, and not unlike the chattering of the fern-owl, or goat-sucker, but more inward.

When mole-crickets fly they move "*cursu undoso*", rising and falling in curves. In different parts of this kingdom people call them fen-crickets, churr-worms, eve-churrs, all very apposite names."

Unfortunately the Mole-cricket is now one of the rarest of all native Orthoptera and is most unlikely to be seen. It has become decidedly rarer over the last 200 years to the point of near extinction in this country, and no existing colony is now known, although the odd singleton is reported from time to time. However it is one of the most elusive of insects, spending almost the whole of its life underground and colonies probably still survive somewhere, quite possibly in Surrey.

Indeed, Surrey probably has as many records of the Mole-cricket as any other county, including Hampshire and Dorset, and therefore all of these are referred to below in chronological order.

The first record, in about 1830, is in *The Letters of Rusticus on Natural History* edited by Edward Newman and published in about 1849. This book is a collection of papers, articles and notes by Newman, a Godalming naturalist, which had appeared previously in magazines, and it contains a fascinating account of the natural history of south-west Surrey at the beginning of the nineteenth century. In a description of nightjars on the Pudmoors of Thursley Common he says "I have seen the nightjar on a turf-stack with its throat nearly touching the turf, and its tail elevated, and have heard it in this situation utter its call, which resembles the birr of the mole-cricket, an insect very abundant in this neighbourhood. I have almost been induced to think this noise serves as a decoy to the male mole-cricket, this being occasionally found in the craw of these birds when shot." He then goes on to liken the call of the insect to "the noise of an auger boring oak, continued, not broken off as is the auger".

Stephens (1835) gives the next record – "at Ripley". This locality is only about one kilometre from a later well-known colony at Send, and probably refers to the same locality.

Burr (1902) did not know of these records, saying that it was unknown in the county but adding that it should be sought for and "may be found in warm sandy commons and in moist spots where the soft ground permits it to make its well-known burrows".

Lucas (1920) records that G.Dalgliesh, who lived at Milford, found a Mole-cricket at Churt in 1901 and another in 1908 and also "one which flew against Mr. Dalgliesh's face outside his garden-gate at Milford about 9 p.m. on 3rd June 1902, and fell down; it was a very fine specimen". This is the only report that I know of anyone in this country seeing a Mole-cricket fly, apart from Gilbert White who said "they move *cursu undoso*, rising and falling in curves". In the Charterhouse School Museum at Godalming there are two specimens of Mole-crickets which unfortunately have no data labels but appear to have been collected by O.H.Latter, who was a master at Charterhouse, and who was collecting in the area in about 1915. It seems possible that these specimens came from the Milford or Thursley district.

Lucas (1921) reported two Mole-crickets in cottage gardens at Send in 1920, one as early as 17 March. There are three specimens in the Fox-Wilson Collection at R.H.S.Wisley, one from a cottage garden at Send dated 7 April 1915, another from the same dated 16 March 1920 and yet another dated 4 August 1936. It was possibly quite common at Send at this time.

Lucas (1928) also reported a nymph found on the hall floor of Tiffin Girls School, Richmond Road, Kingston upon Thames, on 21 December 1927. This is a most surprising date and it is a mystery how it could have got there as nymphs cannot fly. It is possible that it had been introduced with vegetables or plants from abroad, although there is a very recent record from Kingston upon Thames.

There is a specimen in the Natural History Museum, London, which was collected by B.Vesey-Fitzgerald in May 1948 at Wrecclesham near Farnham which is by the River Wey.

The next record was in July 1951 when Mole-crickets were reported to the Plant Pathology Laboratory to be damaging a potato crop at Tilford (Ragge, 1955). This locality consisted of gardens running down towards the river Wey behind the Donkey Inn at Charleshill and the colony apparently existed for some years and may still exist. Chris Haes actually heard one churring there at dusk on 3 June 1954 (Haes, in litt.). He may be one of the few people who have heard the churring of a Mole-cricket in this country. David Ragge visited the site in 1964 and saw unmistakable signs of Mole-crickets in the vegetable gardens, where they had been damaging onion plants (Ragge, in litt.). Ray Fry, one-time warden of Thursley Common, was given a dead specimen at the Donkey Inn in about 1959 by someone who had found it in a potato patch nearby. It was clearly still a familiar insect in the neighbourhood because the locals still knew it by the name of "jar-bob" or "jar-worm". In May 1996 I visited the site with Mike Edwards and ran moth-traps there with the hope of catching a Mole-cricket. Whilst there we spoke to the lady who still owned the vegetable garden who told us that her gardener used to turn up Mole-crickets regularly in the damp seepage in the garden and she remembered him showing one to her in 1964. She also described what was clearly a cast skin of a Mole-cricket, turned up later when digging by the seepage.

In 1953 Mole-crickets were again reported as damaging crops, this time at A.W.Secrett's market garden at Cartbridge, Send (Haes, in litt.). Chris Haes saw one there beside an irrigation pond in May 1953. Unfortunately the 30-acre farm has now been extracted for gravel but it is possible that they are surviving in the gardens of surrounding houses. This is the fourth record for Send in this century, and it seems very probable that the Mole-cricket could still be surviving in the area as there are still plenty of damp flood-meadows along the River Wey. Indeed as recently as 1998 Howard Inns was searching the area and was told by the owners of Highfield Farm, just below Cricket Hill and adjoining Secrett's farm, that when they moved there about 25 years previously, they had been told by the previous owners that there had been Mole-crickets at the end of the garden. It seems quite likely that Cricket Hill is so-called because it was next to a noisy colony of Mole-crickets. There is another Crickets Hill at Weybridge, just north of Brooklands and by the River Wey, which may also have had a colony nearby; the area around would be worth a search. There is yet another Cricket's Hill just north of Hascombe which has seepages on the slopes running down to a stream, and a fourth one, called Cricket Hill, at South Nutfield, again near a stream but I have not yet visited it. Readers may well know of other Cricket Hills in the county and I would be pleased to hear of them in case they give a clue to finding this species. It is possible that the name is derived from the Anglo-Saxon word "clicket", meaning little.

In 1961, and for a year or two after, there were records from Balchins Lane, Westcott near Dorking. They were recorded in gardens there, especially in a garden by a lake, where a drowned nymph was seen in 1962 (Haes, in litt.). Millbridge, near Frensham, again by the River Wey, was another area where Mole-crickets were known at this time. Mrs. Town-Jones lived for some 30 years in a house in this village, which had a seepage in the south-sloping garden. She was a keen gardener who used to find Mole-crickets every year, often in the evenings when watering the frames; she also used to hear them calling regularly in early summer right up to 1979 when she moved house (per ME). Mike Edwards visited the site in 1996 but the garden has now been levelled and there is a tennis court where the seepage was, and there were no confirmed signs of Mole-crickets.

Finally on 15 August 1989 a Mr. Peter Miles unearthed a Mole-cricket while digging in the garden of a private house on Kingston Hill, Kingston upon Thames (Martin, 1990). It may possibly have been introduced with plants from southern Europe, but there had been a previous record from nearby.

All the above records, except for those from Westcott and Kingston upon Thames, are from the Wey valley and there seems no reason why the Mole-cricket should not still be there, as there are still plenty of gardens and allotments bordering the mostly undisturbed flood-meadows.

The Mole-cricket is very difficult to find because it is virtually subterranean and unfortunately in this country it very seldom calls. I have heard a very similar species, the Vineyard Mole-cricket (*Gryllotalpa vineae*), calling from its burrow in Portugal and the noise was very loud and audible from at least 500 metres. The call of the Mole-cricket is much softer and therefore less audible. However it does make raised ridges where its burrowing has been only just below the surface, like a mole, and these are very distinctive, as are its calling burrows – 3 or 4 holes leading down to the central calling chamber.

C.B.Pickard suggests an interesting method of finding them, if a known locality can be found, "by covering parts of the ground around their haunts, after there has been heavy rain that is likely to be followed by a dry sunny period, with sacking or sheets of tin or wood. As the surrounding earth becomes drier the insects are likely to collect under such shelters where the earth will remain moist" (Pickard, 1954). I have certainly heard of people who have found them by turning over planks of wood in Guernsey and in Greece, and Ragge describes finding them in Wiltshire by turning over turves which had been piled up.

Because of the rapid decline of the Mole-cricket English Nature included this species in its Species Recovery Programme in 1994. Mike Edwards was in charge of this project from the beginning and collected data, both historical and current, from all over the country in order to try to build up an overall picture of the distribution and preferred habitat. But it is difficult to study a species when no colony is known to exist. One area in which he hopes to find a colony is the valley of the River Wey and Howard Inns has been helping in this search since 1995. As part of this search he has been looking for suitable habitat, listening for calling males on suitable warm nights, laying roofing felt, concrete slabs and corrugated iron sheets in the hope of finding crickets underneath, and even using the local press and radio in an attempt to elicit reports from the general public and local naturalists. He has paid particular attention to allotments adjoining the flood-meadows, putting up notices and speaking to allotment gardeners. So far this has failed to produce results but the project is continuing and both of them are confident of finding a colony in Surrey before long.

TETRIGIDAE – Groundhoppers

Groundhoppers are small, rather inconspicuous relatives of the grasshoppers, with short antennae but having a very long pronotum which extends over the back of the abdomen and sometimes even further. They have cryptic colouring to match the bare earth on which they are usually found, and they produce no song. They feed on moss and algae, and overwinter as either nymphs or adults. There are three species in the British Isles but only two are known from Surrey, the third one being almost confined to the coast.

Tetrix subulata (Linnaeus, 1758)   PLATE 9   Slender Groundhopper

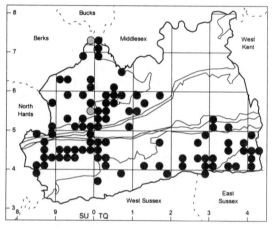

The Slender Groundhopper has a narrow, elongate appearance because of its normally very long pronotum which extends beyond the tip of the abdomen. It has fully developed wings, extending beyond the pronotum, and can fly well; there is a short-winged form with a shortened pronotum. It is usually brown, grey or mottled, and occasionally nearly white.

This groundhopper has a more restricted habitat than the Common Groundhopper, as it only occurs on bare mud in wet or very damp places, such as ditches, pond margins and wet rushy fields. Because of its habitat, it is more localised than its relative, but it is easier to find since it flies in a slight curve for short distances when disturbed. It overwinters as as an adult or a nymph. It is widely distributed and reasonably common in southern and central Britain.

From the map it can be seen that in Surrey this groundhopper is common all along the Wey valley and in wet fields and ditches nearby, and also in the Thames valley in the north-west, as well as the northern Mole valley. It also occurs in the south-east on the weald clay, in damp rushy patches in fields, beside ponds and along ditches. It is absent from the western heaths and bogs, which seem to be too acid, and from the chalk which is too dry for it. There are few records from most of the London clay, especially from close to and in London, and this may also be because it is too acid; for instance it has never been found in Richmond Park which looks superficially suitable.

Being a resident of really wet places, it is adept at swimming. We have often observed that after flying off when disturbed, it frequently lands on water, but quickly swims back to land. Ditches may be an overlooked habitat; otherwise it is difficult to explain how a last-instar nymph came into suburban Horley and climbed a flight of six stone steps to present itself at the door of an orthopterist at 9.30 in the morning (RDH). There are many records of adults in September and a few for nymphs, since some nymphs go through the winter before reaching maturity; they can easily be reared if placed in a container with some earth and moss, and kept in a cool place. Most other records are of adults in April, May and June, extending to the first week in July, but it is difficult to judge whether adults found on three separate occasions in late July and early August (26 July 1987, 7 August 1993, 9 August 1998) belong to the overwintered population or to the new generation.

A very similar species, Cepero's Groundhopper (*Tetrix ceperoi*) is found on the south coast and extends inland in some other counties. Only a small proportion of our records is based on specimens that have been collected and checked, but those that have been checked are all certainly Slender Groundhopper. The remaining field records are all from typical habitats for this species and it is most unlikely that any colonies of Cepero's Groundhopper have been overlooked.

Burr (1902) merely says that it is less common than the Common Groundhopper and is found at Dormans. Lucas (1920) gave no further localities and considered it to be an uncommon insect, but probably overlooked. Its status is probably unchanged, but modern recorders have perhaps found it more often by searching its rather specialised habitat which is not used by other species of Orthoptera. The best method of finding it is by examining bare mud, where it is very wet and not too acid, and looking for the short curving flight when the insect is disturbed.

Tetrix undulata (Sowerby, 1806)   PLATE 9   Common Groundhopper

The Common Groundhopper is short and stubby and is variously coloured grey, brown or whitish and often mottled. It has a high central ridge on the pronotum, which is shorter than in the Slender Groundhopper, and cannot fly because it only has very short wings. Although common and widespread this little groundhopper is difficult to find because of its size and effective camouflage, and also because it makes no call. On the other hand, because it overwinters as an adult or a well-grown nymph,

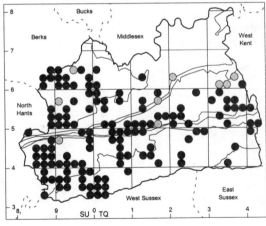

and is also long-lived, it can be found in all months of the year, right through from spring until the end of November. However, in late July and August, most (if not all) specimens found are nymphs. It can even be found in leaf litter during the winter.

In Surrey it is reasonably common and widespread but not apparently in the north-east. There is plenty of suitable habitat in Richmond Park but it has not yet been found there. Richard Jones has observed that it does not seem to like built-up areas, possibly because there is too much disturbance; he has surveyed 25 sites in the London area without finding any, and is sure that the absence from there is real. It occurs on bare soil in three main different habitats: woodland rides and clearings, especially in forestry plantations; chalk downland where the insect is usually very pale and therefore difficult to see on the bare chalk; and heathland where it occurs on both dry and wet patches of bare sand, especially where moss is present. It is particularly attracted to burnt areas and old bonfire sites. Although it is usually found in dry habitats, it does occasionally occur in places where the Slender Groundhopper might be expected, such as by a pond at Earlswood or in the damp grassland of Blindley Heath, where it occurs with the Lesser Marsh Grasshopper (RDH, 1980).

It is commonest on the heaths in the west and on the North Downs but it can occur almost anywhere. Because it is difficult to find, it is probably somewhat commoner than the records show. Although flightless, it seems to be able to colonise new habitat. When a large clearing was made, by coppicing, in the centre of Glover's Wood, the Common Groundhopper was found there in the following season. It persisted for about five years until the trees grew up again (RDH).

Burr (1897) gives Surbiton, amongst other localities, but it has not been recorded from there for many years. The nearest recent record to London was West Ewell in 1969. Lucas (1920) gives many localities. Payne (1958) gives Riddlesdown and although it was not recorded there during an intensive survey of the common in 1995/6, it is still present in the adjacent chalk quarry (RDH). The distribution map shows shaded dots along the north-east edge of its range, which probably indicates a retraction in range, possibly due to human disturbance.

Searching bare ground and leaf litter in likely spots is really the only way of finding it, and, because it is so well camouflaged, it is rarely seen until it hops. The earliest date in the spring on which it has been observed is as a well-grown nymph in leaf litter on 31 March, and the latest date is 21 November (DWB).

ACRIDIDAE – Grasshoppers

The grasshoppers are narrow and compact, with short thickish antennae and strong hind legs which help them to jump long distances. They are variously coloured with distinctive and sometimes complex markings, especially on the pronotum, which can be very useful in identification. All except one species have fully-developed wings and can fly, but normally only over short distances. They are strictly diurnal sun-loving insects, ovipositing in the ground or at the base of vegetation, and producing characteristic songs by rubbing their hind legs, bearing a row of stridulatory pegs, across a prominent vein on their forewings. They all feed mainly on grasses. Females are larger than males. All but one of the ten species which occur on the British mainland are to be found in Surrey, some abundantly.

Stethophyma grossum (Linnaeus, 1758) PLATE 10 Large Marsh Grasshopper

**National Status:
RDB2 (Vulnerable)**

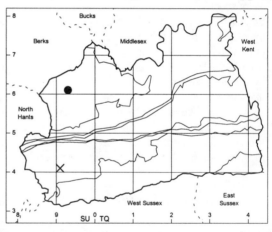

The Large Marsh Grasshopper is the largest and most striking grasshopper in Britain but also one of the rarest, although still widespread in western Ireland. It is now restricted to only a few quaking bogs in southern England, especially in the New Forest and east Dorset. This magnificent insect is also the most colourful of the grasshoppers, being vivid greenish yellow or olive brown, with red under the hind femora; the hind tibiae have black and yellow bands and there is a distinctive yellow line along the front margin of the forewing. The males make their song by flicking the hind tibia against the forewing, a method which is almost unique amongst grasshoppers. The noise is a distinctive ticking repeated about eight times. Geoffrey Collins has observed that they can also stridulate in the normal way, but produce only a brief tuneless buzz.

There appears to be only one native colony of this rare grasshopper in Surrey and this was only discovered in 1982 at West End Common, Lightwater, and has hardly been seen since. There was also a well-documented colony, introduced onto Thursley Common with the authority of the then Nature Conservancy, which may still exist.

Stephens (1835) stated that it was "to be found in marshy districts about Camberwell, etc, near London". At that time the south bank of the Thames was still mostly marshland with villages and was still suitable habitat for this grasshopper, but it must have become extinct there shortly afterwards with the growth of London.

In 1967 it was deliberately introduced with the authority of the Nature Conservancy to an area of quaking bog just west of the Moat Pond on Thursley Common. This small colony persisted until at least 1991 but never spread from its original site and was still only to be found in an area of about 20 square metres. I saw and heard it there in the 1970s but it was apparently not seen or heard, in spite of many searches, from 1979 until September 1991 when Ray Fry and Jonty Denton saw two and heard six in exactly the same spot. This spot has now become drier, like much of the common, and is no longer very suitable habitat. It was searched in 1997 without success (DWB).

David Ragge and I had suspected for many years that it might occur on the quaking bogs of Colony Bog, Strawberry Bottom and Hagthorn Bog near West End and we both searched this latter area in the 1950s. It is however very difficult terrain as most of it is Ministry of Defence firing ranges and strictly "out of bounds". The habitat is almost identical to the quaking bogs in the New Forest where the grasshopper abounds.

However on 14 August 1982 T. Price reported finding one female at Folly Bog, near Lightwater, adjoining Hagthorn Bog, which both David Ragge and I had searched previously. More specimens were found there in 1985. I have searched there twice since then without finding it, but it could still be there in low numbers and it could also occur further south in the closed area. It has been suggested that this colony may be the result of an unrecorded introduction but this seems unlikely. In 1998 Valerie Brown obtained special permission from M.O.D. to enter the closed area immediately to the south and found plenty of ideal habitat consisting of really wet quaking bogs with bog myrtle and bog asphodel, similar to that in the New Forest, but failed to find the Large Marsh Grasshopper. However it was a cold day and she intends to make another search in 1999 as she is confident of finding a colony there. Any further information on this site would be most welcome. This species should most certainly be searched for more thoroughly as it is so rare. It has now been included in English Nature's Species Recovery Programme.

Stenobothrus lineatus (Panzer, 1796) PLATE 10 Stripe-winged Grasshopper

The Stripe-winged Grasshopper is a medium-sized, mainly green grasshopper, with red on the abdomen in mature adults, which separates it from the rather similar Common Green Grasshopper. The female usually has a distinctive white line (the *linea scapularis*) along the lower edge of the forewing and a whitish, comma-shaped stigma on the wing. The song is a pulsating, high-pitched sound quite unlike that of any other grasshopper. It is a rather local grasshopper with a scattered distribution in southern

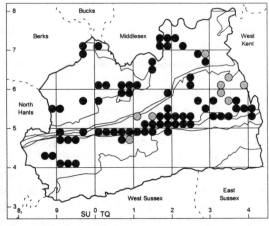

England, confined almost entirely to chalk and limestone but also occurring very locally on sandy soils, mainly in Surrey and northern Hampshire.

In Surrey it is abundant in open downland turf, mainly on the steep south-facing slopes along the North Downs. Here it prefers the warmest sites, like the Rufous Grasshopper with which it often occurs, but it needs short turf with bare patches of soil whereas the Rufous Grasshopper thrives in longer, coarser grass in scrubby areas. But it also occurs in isolated colonies on sandy commons and heaths, where there is plenty of short grass, and Surrey is one of the few areas in England where it is as common off the chalk as on it. Some of the best places in which to find it, in these grass-heath habitats, are on golf courses, where it often occurs at the edge of the "rough" where heather is mixed with sparse tufts of wavy hair-grass; this area between the fairway and the rough is kept reasonably short by occasional mowing and there are usually bare patches. It has also been found in a garden at Ash Vale by Judith Marshall and on a roadside embankment at Runnymede.

It is common or abundant along the North Downs from Compton on the Hog's Back to South Hawke near the Kent border. Off the chalk it had only been recorded prior to the survey at Wisley Common (where it still occurs), Bookham Common, where it has not been seen during the survey, possibly due to scrub invasion, and on the golf course at Mitcham Common in 1957. However in 1975 I started to find more sites on sandy heaths and commons and then on golf courses, so it is clearly widespread but local off the chalk, often in isolated colonies, although it does not occur on the weald clay in the south.

It is widespread on the drier London commons such as Richmond Park and Ham Common and even occurs as near to Central London as Wimbledon Common, Morden Hall Park (DE) and Mitcham Common. It is also on sandy grass-heath commons at Witley Common, Ockley Common, Ash Ranges, Epsom Common, Westfield Common near Woking, Ranmore Common, Albury Heath and Windsor Great Park. It used to occur at Croham Hurst golf course (on sands overlying chalk) prior to 1950.

West Surrey is particularly well-endowed with heathland golf courses and it has been found on the following golf courses – Hankley Common, West Hill near Brookwood, New Zealand at Byfleet, Puttenham, Woking, Esher (all DWB), Betchworth Park and Mitcham Common. There are a few other golf courses with suitable habitat where it could occur.

Both Burr (1902) and Lucas (1920) give various localities on the North Downs from Merrow Downs to Redhill. Payne (1958) mentions Bookham Common and Wisley Common, both off the chalk. It seems that earlier recorders had failed to find the other colonies on the heaths and golf courses. Possibly the creation of more golf courses produced suitable habitat into which it has spread. But it is difficult to explain why it was not found in such well-worked areas as Richmond Park and Wimbledon Common.

It is flourishing on those parts of the North Downs where the National Trust has introduced conservation grazing (KNAA), but the equally impressive grazing operations of the Corporation of London and the Downlands Management Project seem to have come too late for this species; it was entirely absent from the former's commons in the Coulsdon area in 1995/6 (RDH). Our map shows that there was already a significant decline in the north-east before 1970. An earlier mowing regime on many of the sites may have kept the turf short and preserved much of the downland flora, but caused the loss of some of the insects such as this grasshopper. Its long-term survival is also doubtful on the remaining open areas of downland that are maintained temporarily by rabbits or conservation volunteers.

It is possibly a species which is spreading off the chalk but it may be decreasing on the chalk as many of the best short-turf sites are disappearing under encroaching scrub; these sites need to be conserved urgently not only for this grasshopper but for the many other rare or scarce insects and plants which they hold.

Early Date. One heard stridulating on the North Downs near Dorking on 28 May 1990 (MJS).

Late Dates. Two females at Pickett's Hole above Wotton on 6 November 1975, and a female on Witley Common on 2 November 1975 (both DWB).

Omocestus rufipes (Zetterstedt, 1821)   PLATE 10   Woodland Grasshopper

Nationally Scarce B

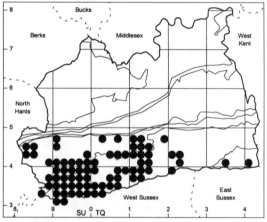

The Woodland Grasshopper is very dark brown or even black with green, scarlet, black and white markings on the abdomen and sometimes, in females, green on the pronotum and tops of the forewings. Some specimens look very like their close relative the Common Green Grasshopper; their song is also similar to that of the Common Green Grasshopper but of shorter duration. The best distinguishing characters are the conspicuous, chalk-white palps which show up against the black underside of the head and, in mature adults, the bright scarlet tip to the abdomen; it is usually necessary to catch one and turn it upside down to see these characters. This attractive grasshopper is very local in southern England with its main concentration in the Weald and the New Forest. There are very few records north of the Thames although it has recently been found in East Suffolk.

In Surrey this grasshopper is also local but can be common in its favoured haunts. As its name implies, it is normally found in sunny woodland rides or clearings or where woodland has been felled and there is sparse vegetation for a few years, but it does also occur occasionally on heathland, both damp and dry. It has not yet been found on chalk during the survey period but there is an old record from Box Hill (Burr 1902).

Except for a few localities in the south-east, it is confined to the south-west of the county, In this area however it is locally common to abundant. It is commonest in the woods and forestry plantations on the weald clay around Chiddingfold, Dunsfold and Cranleigh, but it is also reasonably common on the greensand hills around Haslemere and Tilford, and on Winterfold, Hurtwood and Leith Hill, especially on the sunny south-facing slopes. It has been seen occasionally on damp heathland on Thursley Common and on dry heathland on Hindhead Common, but near to trees (DWB). In the south-east it occurs on the weald clay at Glover's Wood near Charlwood, at Newdigate, Hedgecourt and Dormansland. The latter sites are outliers from a more extensive distribution in the High Weald of Sussex. It has not been found in the north of the county, but it does occur in Windsor Great Park in Berkshire, so may possibly yet be found in the north-west.

Burr (1902) cites a record for Box Hill, and there is an unconfirmed record from there in 1961. I had always considered these to be misidentifications as many old records of this species from outlying localities have been proved to be misidentified Common Green Grasshoppers, but the species has since been found in scrub and woodland rides on chalk in Sussex, so the record is not impossible. There is also a 1937 record for Oxshott in the north-west which may have been a misidentification.

Burr (1902) only mentions the Box Hill record, and Lucas (1920) only mentions Leith Hill. They had obviously either overlooked the other sites or it was not present then. Other authors have suggested that it has only spread very recently, owing to the recent increase in forestry plantations and their rotational management, which provides ideal habitat for it. It seems inconceivable that older collectors could have failed to find it in the Weald where it is now so common, except that they never apparently visited the wealden clay area around Chiddingfold and Dunsfold which is now its main centre. It is possibly an overlooked species, being perhaps mistaken for the Common Green Grasshopper.

Searching short grass and bare areas in sunny woodland rides and clearings is the best way of finding this grasshopper. The scarlet tip to the abdomen is very distinctive when it flies off, but a specimen should be caught and examined in the hand before a positive identification is made. Look for the white palps and scarlet tip to the abdomen on the underside, and beware of the similarity of its song to that of the Common Green Grasshopper.

Late Date. One was seen on 6 November 1977 at Lythe Hill, Haslemere (DWB).

Omocestus viridulus (Linnaeus, 1758) PLATE 11 Common Green Grasshopper

As its name implies, the Common Green Grasshopper is one of the commonest and most widespread grasshoppers in the British Isles. It is medium-sized, the female being rather large, and is usually green, at least above, and rarely purple on the sides and hind femora; the male may be green or brownish. Neither sex has any red or orange on the abdomen. The side-keels on the pronotum are gently incurved. The song is a continuous mechanical churring lasting for up to 20

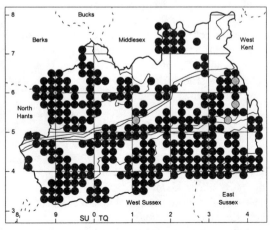

seconds, starting softly but becoming louder. It is found wherever there is undisturbed grassland in moist places, even high on hills.

However in Surrey it is widespread but by no means common, being absent from many areas, possibly because Surrey is generally too dry for its liking. It prefers undisturbed long grass and damp meadows and is not therefore found on the dry sandy heaths in the west, except at the edges where it occurs frequently under bracken. It is however frequent on the grassy edges of damper heath with bilberry on the moist greensand hills such as Leith Hill. It is also found on the chalk if there is undisturbed long grass, and is thriving on Farthing Downs where the Corporation of London has introduced low-intensity cattle grazing, but it does not occur in built-up areas. It is however very common in Richmond Park and is found as near to central London as Putney Heath, Wimbledon Common, Mitcham Common and Wandsworth Common, where there is still some undisturbed damp grassland.

It is probably commonest on the clay in the south where it occurs in open grass and also in woodland rides and clearings, if there is plenty of grass; it is particularly numerous in the wealden clay woods around Chiddingfold and Dunsfold, e.g. Botany Bay and Sidney Wood, where it occurs with its close relative the Woodland Grasshopper.

It is one of the earliest grasshoppers to mature and it was heard stridulating on 21 May 1990 on the North Downs near Dorking (MJS). The first dates for adults are generally in late June or early July.

A one-legged specimen was observed calling from a bramble patch at Horley (RDH). Perhaps it is normal for grasshoppers to use one leg at a time.

All the old authors describe it as common and widespread and its status has probably not changed.

Chorthippus brunneus (Thunberg, 1815)   PLATE 11   Field Grasshopper

The Field Grasshopper is medium-sized and variable in colour, being normally brownish but sometimes mottled or striped and occasionally purple or rarely green; the tip of the abdomen is usually orange or red in mature males. The underside of the thorax is distinctly hairy and the long wings, with a distinctive bulge on the lower margin of the forewings, enable it to fly well. Its best distinguishing characters are the shape of the pronotum, the side-keels being sharply incurved, and the hairy underside of the thorax.

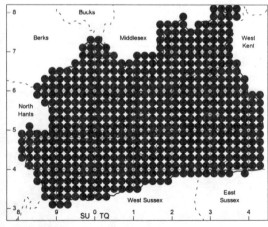

The song is a series of short chirps. It is widespread and very common in the British Isles but most abundant in southern England.

In Surrey this is certainly the most widespread and abundant of the grasshoppers, especially on sandy soil where there is bare ground with a little grass. The reason for this is that it flies readily, especially in hot summers, and can therefore colonise the smallest piece of suitable habitat even in the heart of large cities such as London. In the recent long hot summers, when numbers have been high, I have noticed how frequently these grasshoppers, when disturbed, will not only hop large distances with assisted flight, but will just keep on flying for 50 metres or so. On one occasion I saw one rise to about 20 metres and fly across the River Mole, and it was still flying at this height when it disappeared from view, through binoculars, at a distance of about 100 metres.

It prefers dry sunny places on sand or chalk with short grass and is abundant on rather bare roadsides, but is absent from really wet bogs and marshes, heavy clay, regularly mown grassland and thick woodland; it is usually absent or rare on dry heathland unless there is at least some grass with the heather. It oviposits only in dry soil and is therefore scarce in damp sites, although there is nearly always an anthill or molehill which is dry enough for the purpose.

As will be seen from the map, it has been found in every tetrad in the county. It is doubtful if any other county can equal this. It was easier to find in Inner London than on the damp, flat, heavily agricultural clay of the south-east, where it is scarce and is only to be found on warm south-facing banks. Almost every churchyard, cemetery, abandoned building site, allotment, park or common in Inner London had at least a few square metres of suitable habitat and there inevitably this grasshopper could be found. Unfortunately the local councils tend to mow or spray almost every square metre of their parks and cemeteries these days, but usually a small corner or edge of natural grass can be found. Even in New Cross, which has virtually no open spaces, I found it on a piece of waste ground about 2 metres square which had a few clumps of grass, at New Cross Station. Other localities in Inner London are Battersea Park, Hammmersmith towpath, The Cut at Waterloo, Vauxhall, Southwark, Surrey Docks at Rotherhithe, Camberwell Old Cemetery, New Cross Cutting and Brockwell Park in Brixton.

A form with some green colour is frequent in Surrey and may be confused with the Common Green and Stripe-winged Grasshoppers. The structural characters given above should be carefully checked. The very rare form with head and pronotum entirely green was found at Horsell Common in 1967 by J.A.Meadows (Ragge, 1973).

This was always a common insect in the past and is still abundant in most places.

Walking through suitable habitat usually puts them up as they are very easily disturbed. They hop or fly much further on the whole than other species; their song is also loud and distinctive. Two males are often found calling in alternation and the human observer may participate in such a rivalry duet by making a short, high, hissing sound, to which an alert grasshopper will generally respond – a rare example of conversing with an insect (RDH).

First dates for adults are generally in July but can be as early as 15 June.

Late date. A male on 3 December 1975 at Newlands Corner.

Chorthippus parallelus (Zetterstedt, 1821) PLATE 12 **Meadow Grasshopper**

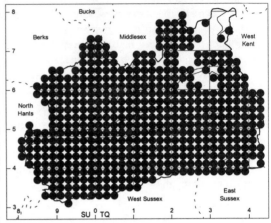

The Meadow Grasshopper is medium-sized and mainly green, although other colour forms such as brown, or even purple in females, do occur. Males are much smaller than females and have forewings reaching to the end of the abdomen, whereas females have only very short forewings and look like overgrown nymphs; both sexes have only vestigial hindwings and cannot fly, although there is an uncommon long-winged form which does fly. The best distinguishing character is the pattern of the pronotum, the side-keels being not quite parallel but gently incurved. The male's song is a short burst of rasping sound repeated every ten seconds or so. It is the most widespread grasshopper in Britain and also probably the commonest, occurring in all sorts of grassland habitats.

In Surrey this is at least the second most common grasshopper, after the Field Grasshopper. However its habitat is rather different as it prefers rather damp green grass (because it oviposits only in moist soil whereas the Field Grasshopper oviposits in dry soil, although the two often occur together) and is therefore difficult to find in Inner London. It does seem to be able to tolerate more disturbance, such as regular mowing, and can therefore survive on road verges and fields sown for silage. It is often the only grasshopper among the coarse grass of roadside verges, where it can be abundant; 50 specimens were sitting on a single drain-cover in a roadside verge at Sidlow on 9 September 1977. It is also found commonly in wet heathland and bogs, where the only other orthopteran is usually the Bog Bush-cricket. It even occurs on dry sandy soils so long as there are some moist grassy areas, and also in woodland rides; one was seen on a bramble leaf two metres up in a hedge (RDH).

There is a macropterous (long-winged) form *explicatus* (Selys) which, though not common, is regarded by Geoffrey Collins as being probably present almost every year in colonies of any size. He kept fairly regular observation on a colony at Bramley Bank, Selsdon, which produced a few of this form for 10 out of the 12 consecutive years 1979-1990, and has numerous other records of scattered occurrences over the last 50 years.

It seems probable that the long-winged form occurs when there is a large build-up of numbers and the colony becomes overcrowded. On 5th August 1996 Ian Menzies was on Epsom Common when he saw considerable numbers of insects flying slowly in the hot sunshine three to twelve metres above a field of long grass to the east of the Christchurch car park. On investigation he was amazed to find that these were long-winged Meadow Grasshoppers; the normal form of the species was hyper-abundant in the area at the time. He considered that this may have represented a generation of macropterous individuals with a potential for dispersal. This long-winged form can be mistaken for the Lesser Marsh Grasshopper.

This species is abundant and widespread except in heavily built-up areas. The only records near to central London are from Nunhead Cemetery, Peckham, and Battersea Park, but it could be in other private cemeteries or on allotments or railway land.

All the old authors describe it as abundant and widespread, although not in Inner London. Its status has probably not changed. Because it cannot fly, it is normally incapable of recolonising isolated areas, such as the commons of Inner London, except for the long-winged form which is capable of long-distance flights.

Walking through any grassy area will almost certainly turn it up, and being flightless, it will only hop a short distance and can be easily caught for identification. The song is normally unmistakable, but in cool weather, or towards evening, it slows to half its normal rate and can then sound quite different.

Early Date. An adult on 10 June 1990.

Late Dates. A male at Newlands Corner on 21 November 1975 (DWB) and a nymph at Alfold on 6 October 1991 (DWB).

Chorthippus albomarginatus (De Geer, 1773)   PLATE 12   Lesser Marsh Grasshopper

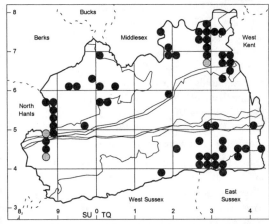

The Lesser Marsh Grasshopper is medium-sized and either pale dull green or straw-coloured with rather short wings which do not reach the end of the abdomen; in spite of this both sexes can fly. The female usually has a distinctive white line on the leading edge of the forewing, which often continues along the edge of the pronotum and the head. The best distinguishing character however is the shape of the side-keels of the pronotum, which are almost parallel. The song is a series of short chirps, each chirp being softer and longer than in the similar song of the Field Grasshopper.

It has a curious distribution in southern England and Ireland, being mainly coastal but having increased and spread considerably inland in the last 30 years so that it is now widespread in inland eastern England. It occurs in a wide range of habitats from saltmarshes and sand dunes to inland marshy areas and flood-meadows and now even urban parks and commons.

In Surrey it also has a strangely patchy distribution, being commonest in damp fields in the south-east and on dry clay commons in London. It also occurs in flood-meadows along the River Wey, and around old flooded gravel pits in the Blackwater valley. It appears to be absent from many localities which appear suitable for it, and it is most difficult to predict its presence in any area.

In 1970 the only known sites were Blindley Heath, Itchingwood Common and Mitcham Common (where there was a small colony in the early 1950s, GBC), all being damp commons on clay in the east. It also occurred in a rather dry meadow at Shirley in 1945-46 (J.A.Whellan). But in 1977/8 Roger Hawkins, who had recently moved to the area, found it in damp rushy fields and commons in the south-east, principally south-east of Horley and north of Lingfield within the areas of the former Horley and Lingfield Commons. The tufted hair-grass was often an indicator of its presence, along with rushes. It seemed to be a "native species" of long-established wet grassland. At the same time I was finding it in small discrete colonies in flood-meadows along the Wey valley around Old Woking, Send, Pyrford and Waverley Abbey, but there were many similar flood-meadows where it could not be found.

Then in 1984 Michael Skelton found it on some of the inner London commons such as Clapham, Tooting Bec and Wandsworth; these commons are quite dry and mostly maintained as urban parks. It was later found in Old Deer Park at Kew and even further into central London at Elm Park, Battersea (DWB); in 1997 it was found in a very dry sandy area of Richmond Park, near Kingston Gate (ISM). It was refound on Mitcham Common in 1998 (DAC, GAC), and in 1990 a flourishing colony was found 3 kms to the east at Heavers Meadow, Selhurst (DWB); this is a small open grassy area of only about one hectare surrounded by buildings, but with a small stream running through the middle. Recently Geoffrey Collins has found it in various sites around South Norwood, including South Norwood Country Park, and Lloyd Park, Croydon, where he had not seen it before. In 1998 he found it on Addington Hills which is a dry area at about 130 metres above sea level. In 1989 Roger Hawkins also first recorded it in the Blackwater valley where it is now widespread in damp fields and around old gravel pits. Away from these main areas it has been found in apparently isolated colonies at Epsom Common, Lightwater, Broadstreet Common near Guildford, near Virginia Water and by the River Hogsmill at Berrylands (but not seen there recently (ISM)). By the 1990s some sites in the south-east had been lost, but then rather more new ones were discovered.

Burr (1902) does not mention it and Lucas (1920) (apart from mentioning Box Hill, an unlikely site) only gives Tilford and Hale. Although it has not been refound in either of these places, it does still occur along the Wey valley between them. Payne (1958) gives Mitcham Common, but it was not then known from the other London commons which would have presumably been well searched.

Why was it hardly known until the present survey started? It was either overlooked or it was very rare and has spread recently. It is a species that is not easily recognised and it tends to occur in discrete colonies, covering only a few square metres, so that it could have been missed by early recorders. However some of the colonies are in well-recorded areas, such as the London commons, and are most unlikely to have been overlooked; Geoffrey Collins has found new colonies in sites which he has known for many years and in which it certainly did not occur previously. It seems clear thefore that there has been a recent expansion of range, especially in and around London; it is difficult to be certain about the area on clay in the south-east or in the Blackwater Valley as both these areas were neglected by earlier recorders.

Late Dates. Still abundant on Blindley Heath on 29 October 1975 and a few at Coleford Bridge Gravel Pit, Farnborough, on 23 October 1990 (DWB).

Gomphocerippus rufus (Linnaeus, 1758)   PLATE 13   Rufous Grasshopper

Nationally Scarce B

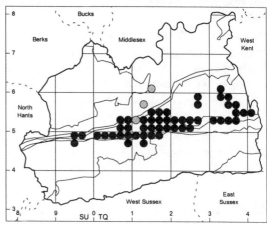

The Rufous Grasshopper is predominantly dull brown in colour, with pale orange on the end of the abdomen in mature adults; occasional females are reddish purple. Its most distinctive character is the white-tipped, clubbed antennae, quite unlike those of any other grasshopper. These are twirled around by the male in its extraordinary courtship dance. The normal song is a soft, buzzing chirp lasting about five seconds. It is a very local insect, being restricted to warm grassland sites on chalk and limestone in southern England, usually under or near trees and scrub.

In Surrey this interesting and very localised grasshopper is as common on the North Downs as anywhere in England. Here it is abundant from the Hog's Back above Puttenham in the west to the Kent border in the east, occurring mainly along the steep scarp slope but also on south-facing slopes to the north such as Riddlesdown, Chipstead Valley, West Horsley and Leatherhead. The only two small gaps along the scarp slope which appear on our map are at Guildford and Merstham. It prefers warm sheltered south-facing slopes of rough grass but is never far away from trees or scrub. Being a late-maturing species (often being found in November) it has been suggested that it shelters under leaves for protection. It is apparently a woodland species on the continent.

A female of the uncommon reddish-purple variety was seen in 1947 on the Downs above Oxted (GBC), another in 1974 on the Hog's Back at Compton, and another in 1976 on White Downs (DWB).

There are also a few rather curious records from localities away from the chalk. Chris Haes found a small colony at Milford beside the A3 in 1976 (I failed to refind this in 1990). There was a small colony on greensand by the A3 at Compton in 1974, and a female at the edge of a wood at Friday Street on greensand, about 3 kms south of the chalk, in 1976 (DWB). Michael Skelton found it just off the chalk at Gomshall and Dorking and there was a small colony on a south-facing bank at Buckland sand-pits, 1 km from the chalk, in 1981. A singleton on weald clay on a roadside at Parkgate in 1976, about 5 kms south of the chalk, may have been carried from the chalk on a car. It may be that in hot summers, some adults disperse quite a distance and may even start a small colony which exists for a time in unsuitable habitat. Lucas (1920) mentions a well-known colony that apparently persisted for some years on Bookham Common, which is on London clay, and he actually illustrates his book with a painting of the antennae of a specimen from there. Burr (1897) also mentions Oxshott which is many miles from the chalk, but this is more likely to have been a misidentification. Burr (1902) records that there is a specimen in the Dale collection which was collected by Samouelle in the early 19th century from Battersea Fields, but this seems improbable.

Burr, Lucas and Payne state that it was common along the North Downs in the past.

Its present status appears to be unchanged; it still seems to be as abundant at the end of the survey as it was at the start, in spite of scrub encroachment. It was, however, extremely difficult to find at Riddlesdown in the last years of the survey (1994, JP; 1996, RDH).

Searching the steepest slopes of the North Downs and especially warm south-facing hollows in the autumn is the best way to find this grasshopper. The song is distinctive and the insect is very active and the white-tipped, clubbed antennae are easily seen.

Late Dates. 16 November 1978 on the Hog's Back, still abundant; also there on 12 November 1977 (DWB); 11th November 1975 on Colley Hill – 2 males and 2 females (DWB).

Myrmeleotettix maculatus (Thunberg, 1815) PLATE 14 **Mottled Grasshopper**

The Mottled Grasshopper is the smallest of the grasshoppers, the male being especially small; it is also the most variable in colour, being mainly dark, but parts of the body may be green, purple, brown or black giving it a mottled look which blends well with its habitat of broken ground. Its best distinguishing character is the slightly out-turned clubbed antennae (but not white-tipped as in the Rufous Grasshopper), these being more pronounced in the male. Another good character is given by

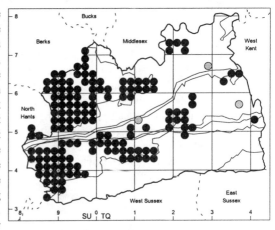

the strongly incurved side-keels on the pronotum, which almost form a cross. The song is a distinctive series of buzzing chirps, starting softly but becoming louder and ending suddenly. It is also the most widespread of the grasshoppers, occurring throughout the British Isles, often abundantly, but only on dry short turf with bare patches of soil in full sun.

In Surrey this attractive little grasshopper is to be found, sometimes abundantly, on almost every dry sandy or heathy common, especially where the grass or heather is short and there are bare patches of soil on which it can sun itself. It is usually the first species to recolonise burnt heathland; specimens on burnt ground are often very dark, almost black, matching the surrounding soil, whereas the normal colour form on the unburnt heathland is green and purple, matching the heather. It does not occur on the weald clay and there are only three records of it on the chalk. It occured in a chalk-pit on Merrow Downs golf course in 1974 (DWB), a locality mentioned by Lucas (1920), at Box Hill (GBC, 1949) and in Brockham Quarry, where it was abundant on the cliff face (DWB, 1994).

It is common and widespread on all the heaths, especially in the west, but it also occurs on some small patches of heath in the east such as parts of Limpsfield Common, Addington Hills and Croham Hurst. It also occurs as near to London as Wimbledon Common, Putney Heath and Richmond Park. It used to be on Mitcham Common, but has not been seen there recently, although there are still small patches of heathery ground on the golf course.

Both Burr (1902) and Lucas (1920) describe it as common on the western heaths and it is probably as common now as it was at the start of the century. Lucas gives two localities on the chalk (Box Hill and Buckland Hill) where it has not been recorded since 1970. He also mentions Merrow Downs, where it was refound. Both authors also mention Blindley Heath, a remarkable record since this is now a damp clay common where the Lesser Marsh Grasshopper flourishes. Perhaps there was some heathland near here in the past, hence the name.

Since it matures early (sometimes by late May), it also dies off early, usually by the end of September, but it sometimes survives into November.

Late Dates. Upper Hale, Farnham, one female on 10 November 1977 (DWB). Witley Common on 4 November 1975 (DWB).

DICTYOPTERA:
BLATTELLIDAE – Cockroaches

Cockroaches are not now included in the order Orthoptera but in a separate order Dictyoptera. There are three species native to Britain and all of these are much smaller and more delicate than the better-known introduced foreign ones. They are very skulking and difficult to find, inhabiting heather litter and grass tussocks, but the males fly and on hot days can be quite active and visible. They are omnivorous scavengers. The Dusky and Tawny Cockroaches have a two-year life-cycle, overwintering as eggs and then again as well-grown nymphs, but the Lesser Cockroach has only a one-year life-cycle and does not overwinter as a nymph. All three species occur in Surrey. Many people seem to be unaware that there any native cockroaches living in the open countryside in Surrey and probably only a few people have actually seen one.

The different species can be distinguished as nymphs (Brown, 1973), but we have not had the confidence to attempt this in the present survey. Our rearings of nymphs were unsuccessful and we never had both common species in captivity at the same time. Consequently a number of additional records, of nymphs alone, have been discarded; all of these, however, fall within the range of the Dusky Cockroach in west Surrey.

Ectobius lapponicus (Linnaeus, 1758)   PLATE 15   Dusky Cockroach

Nationally Scarce B

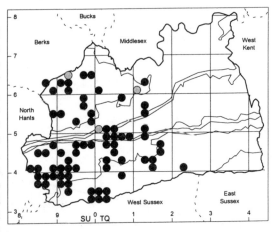

The Dusky Cockroach is small and light brown, the male being almost grey with a darker pronotum, the female brown with a darker underside. The male has long wings, giving it a slender look, whereas the female has shorter wings which do not quite reach the end of the abdomen, giving it a rounder look. Nymphs and females of this species are difficult to separate from those of the Tawny Cockroach. It is found in rough grass and heather in sheltered spots very locally in southern England, being almost restricted to the western Weald and the New Forest.

West Surrey is the main stronghold of this cockroach in Britain; but although it is widespread and common or even locally abundant throughout west Surrey, it is entirely absent from the east of the county, in spite of there being plenty of suitable habitat. The most easterly localities are Glover's Wood at Charlwood and Holmwood Common, both on the weald clay, and Esher Common on the London clay. This insect was first found at Holmwood Common in 1974 and last seen there in 1998, so the eastern limit of its range has remained unaltered throughout the period of this survey. The record for Glover's Wood dates from 1977 and appears somewhat isolated, although an unidentified cockroach was seen on a roadside at an intermediate site in 1984. This distribution is remarkable for cutting across geological features.

In west Surrey it occurs commonly in warm sheltered sites on sandy heaths (the nymphs being often found in large numbers in litter under large clumps of heather), in woodland rides, clearings and edges, and in grassland near thickets either on the chalk downs or on sand or clay. It is often seen on hot days in summer when it is very active in sunshine, the males running about on the grass and heather and flying short distances if disturbed, although Roger Hawkins once noticed that many males on grass stems on Chobham Common became active only when the sun went in at about 4 p.m. They are also frequently beaten out of bushes, but then often run quickly and escape from the beating tray. Males are often attracted to light at night and fly into houses.

Burr (1902) mentions five localities, in all of which it still occurs today. Lucas (1920) repeats these and adds another six, all from west of the Mole, including Oxshott which with Esher Common is the nearest known locality to London. Its status is probably unchanged.

As an adult it is most easily found in June and July on warm sunny days when it is very active. The male runs around on rough vegetation and sits on tops of grass stems and heather, and flies actively when disturbed. During the rest of the season it is best searched for in leaf litter in woods or in heather litter on warm sandy commons. Find a large mature clump of heather and turn up the edge on the south side. Then turn over the leaf litter and the shiny brown, very active nymphs will soon be seen diving for cover. The adult males should be easily identified but the nymphs and females, being difficult to separate from those of the Tawny Cockroach, should be caught for positive identification by an expert. In the winter the nymphs can also be found hiding in the centre of grass tussocks.

Ectobius pallidus (Olivier, 1789) PLATE 15   Tawny Cockroach

Nationally Scarce B

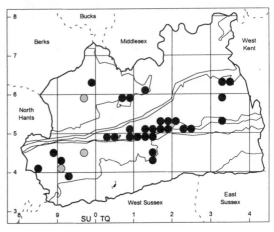

The Tawny Cockroach is small, delicate, golden-brown and fully winged in both sexes. The female looks very similar to the Dusky Cockroach, but it has longer wings which reach to the end of the abdomen. It frequents the same habitat as the Dusky Cockroach but is more often found on chalk downland than elsewhere. Although this cockroach is more widespread in southern England than the Dusky Cockroach, it is on the whole less common.

In Surrey it is also more widespread, occurring right across the county from Frensham in the west to Croham Hurst in the east, but it is nowhere common, or at least easy to find, except on the North Downs between Newlands Corner and Box Hill. The great majority of its localities are on chalk, where it is usually found in grass tussocks on warm, sheltered slopes, but it also occurs on heathland, where it is found in short dry heather, and in woodland rides and clearings, where it is found in leaf litter. Other chalkland localities in the east are Riddlesdown Quarry and Quarry Hangers. Localities where it has been found in dry heather are Thursley Common, Chobham Common (1994, AJP), Wisley Common, Oxshott Heath and Leith Hill; the only woodland record is from Crooksbury Common. The comparatively few dots on our map probably reflect the difficulty which recorders have in finding this little cockroach rather than its true distribution; it is almost certainly more common and widespread.

Burr (1902) only gives Box Hill. There is a specimen in the Charterhouse School Museum which was collected at Godalming by O.H. Latter in 1914. Lucas (1920) thought it occurred only east of the Mole, especially at Box Hill (although an old Woking record is given), and that the Dusky Cockroach occurred only west of the Mole, implying that the two species did not overlap. However the Tawny Cockroach is now recorded in numerous places west of the Mole and both species often occur at the same site, e.g. Newlands Corner.

It was probably under-recorded in the past and although it has now been found in many new sites, its status is unlikely to have changed.

Specimens were occasionally sieved from leaf litter or beaten from bushes at Riddlesdown Quarry, nymphs being found on 15 April, 14 and 19 May, 29 August and 11 September, and adults on 8 August and 11 September (RDH). Some adults therefore survive until September, by which time the over-wintering nymphs of the new generation are already well-grown; the same applies to the Dusky Cockroach.

It is not an easy insect to find, being more nocturnal than the Dusky Cockroach, although sometimes on hot days it can be active, running around on rough vegetation. The best way to find it is by searching grass tussocks and heather and leaf litter, and by beating bushes, in the same way as for the Dusky Cockroach. Both adults and nymphs should be caught for positive identification.

Ectobius panzeri Stephens, 1835   PLATE 15   Lesser Cockroach

Nationally Scarce B

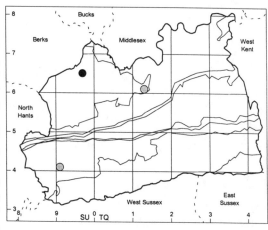

The Lesser Cockroach is very small and dark, the fully-winged male being particularly dark and resembling a small Dusky Cockroach. The female has very short wings, so that the distinctive body segments are clearly visible. The nymphs, which are more likely to be found than the adults, can be easily separated from other cockroach nymphs by the distinctive white margins to the upper segments of the body; they may be seen feeding on pollen in buttercups and other flowers in early summer. Otherwise it is a very skulking insect, easily overlooked.

This cockroach is restricted in Britain to the coasts of southern England and Wales, except for a few inland sites mostly not far from the coast, such as the New Forest. One would not therefore expect to find it in Surrey, but Lucas (1920) does give a record of one collected by G.Dalgliesh at Thursley, presumably on Thursley Common which has the preferred habitat of dry sandy heathland.

It was recorded from Chobham Common in about 1900 (per VKB) and also from Esher in 1946 by T.R.Eagles (per C.H.Andrewes) but it has been suggested that it could have been brought there in sandbags filled with sand from the coast during the war. However there was also a report from nearby Oxshott Heath in 1961.

No specimens from these localities appear to be in existence and I had therefore dismissed them as probable misidentifications. However Professor Valerie Brown, whilst carrying out her long-term study into the life habits of the native cockroaches at Imperial College, Ascot, discovered 6 nymphs and 2 adult females on the southern part of Chobham Common in July 1969. She found more nymphs in subsequent years at the same site and another adult female in 1981. All these specimens were caught by using a suction sampler in litter under heather. In 1983 she discovered another colony in heather on Yateley Common, Hampshire, about one kilometre from the Surrey border. It is clearly very elusive and overlooked. It seems therefore very likely that the older records are correct and a special search needs to be made for this insect in heather litter on dry, hot, preferably south-facing slopes of which there are many in west Surrey. I have searched on Hankley Common, Thursley Common and Frensham Common, but so far without success.

Its habitat is dry heathland and the best method of finding it is probably by searching through the litter under clumps of heather on the sides of tracks or on south-facing slopes. Obviously a suction sampler would be the most effective method, but sieving litter might also work. Any specimen caught should be retained and sent to the Natural History Museum, London.

DERMAPTERA:
LABIIDAE AND FORFICULIDAE – Earwigs

Earwigs are familiar insects which are distinguished by their large pincers or forceps at the end of the abdomen. They are all some shade of brown in colour and are omnivorous scavengers. There are only four native species in Britain and all of these are found in Surrey.

Labia minor (Linnaeus, 1758)   PLATE 16   Lesser Earwig

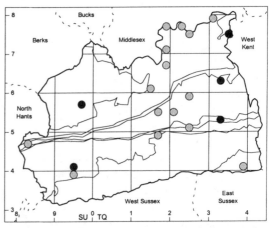

The Lesser Earwig is very small and dull brown with a dark head, looking more like a small rove-beetle or even a large ant than an earwig. It is fully-winged in both sexes and flies readily by day and night and is thus sometimes found in moth-traps. It lives in well-established manure-heaps (preferably of horse manure) and sometimes in compost heaps, and may therefore be a long-established naturalised species rather than a native one. It apparently breeds throughout the year.

It was common in Victorian times when there were many more horses, but it became scarcer as cars and buses replaced them. It is probably now reasonably common and widespread in Britain although, like all earwigs, it is overlooked and therefore under-recorded.

Lucas (1920) gives many localities for this species, including Southwark, and it was probably common and widespread in the early part of this century. But there were then very few records until the 1990s; Lucas reported one flying into a lighted room at Barnes in August 1930. Several, of which one was identified, were found in a moth trap at South Croydon on 30 June 1995 (GAC) and others were found in dung-heaps at Bletchingley in 1995 (P. Kirby) and Milford in 1998 (DWB) and another in a garden at St John's, Woking, in 1996 (AJP).

Richard Jones even saw one flying on a hot sunny day in his garden at Nunhead, Peckham, in 1998, which he caught by hand for identification.

It is probably quite widespread in heaps of horse dung and the best method of finding it is to dig about six inches into the heap where it is moist and warm. The earwig is very active and will probably dive into the manure as soon as it is exposed. Remember that it is very small and as soon as you see it dive, grab the manure and put it onto a sheet of white cloth or paper and then wait for it to crawl out. John Widgery told me of this method and I found my first specimen in the heap of horse dung in my own garden within five minutes, in spite of my previously searching the dung-heap for this earwig intermittently for about twenty years.

Apterygida media (Hagenbach, 1822)   PLATE 16   Short-winged Earwig

Nationally Scarce B

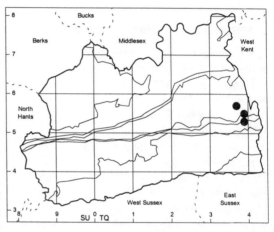

The Short-winged Earwig is smaller than the Common Earwig and very similar to Lesne's Earwig, since it too has no visible wings. The male forceps differ from those of Lesne's Earwig because they are long and gently curved, with only a tiny interior tooth at the mid-point, but it is extremely difficult to separate the females of these two species. John Widgery has recently found some small but significant differences which appear to be consistent. The two main ones are that the Short-winged Earwig is of a far brighter and more uniform reddish colour, and that the shape of the vestigial wings, which are visible through the transparent wing-cases, differs slightly in the two species.

In Britain this earwig is restricted to the south-east corner of England where it is locally common in hedgerows and at woodland edges. It is most common in Kent where it used to occur commonly in hop-gardens and was known as the Hop-garden Earwig. It is a nocturnal species and therefore overlooked and under-recorded, and not much is known about its life history. It is usually found as an adult in August and September and is best located by beating shrubs and hedges or sweeping rough vegetation.

Until 1997 this earwig had not been recorded in Surrey although it was known from Sevenoaks which is only 15 kms from the Surrey border. However J.H.P.Sankey and W.B.Broughton told R.M.Payne that they had found a few specimens in June 1955 under flints on the northern slopes of Box Hill; unfortunately no specimens were preserved (Payne, 1958). This record was not accepted as the location seemed unlikely.

On 18 September 1997 Roger Hawkins discovered one male and four females of the Short-winged Earwig when beating poplar-suckers, nettles and thistles beside a field on the outskirts of Oxted. Luckily he knew the insect from recent entomological trips to northern France and instantly recognised what he had caught. On the same day he found two further specimens on tall vegetation on a roadside verge near the M25 bridge north of Oxted. Later that same day he found another male on low scrub on chalk downland in a dry valley recessed into the North Downs at Woldingham. This constitutes the most westerly locality for this species in Britain. It was found again in October 1998 at the Oxted site by Graham Collins.

It seems therefore that this species is locally resident in the extreme east of the county. It is unlikely to have spread recently from Kent and has probably always been resident in Surrey but overlooked because very few people look for earwigs. It probably occurs in other suitable habitat in the east of the county and is worth looking for in sheltered spots with old hedges and shrubs.

Forficula auricularia Linnaeus, 1758   PLATE 16   Common Earwig

The Common Earwig needs no description as it is one of the most familiar of all insects. It is ubiquitous and common throughout the British Isles, and being fully-winged it can fly, although it is only very rarely seen in flight. It is best distinguished from the other two similar-sized, but rare, earwigs by the pale, folded wings which protrude from beneath the elytra or wing-cases.

Because this earwig is so common very few people bother to record it and no distribution map has therefore been prepared. It can however be safely assumed that it occurs in every tetrad in Surrey; certainly it was found in every tetrad which was searched for it. Any locality found, beyond doubt, without it, would be of great ecological interest. The female is renowned for the care of her eggs and nymphs; in May 1997 Roger Hawkins observed females showing such maternal behaviour on clifftop grassland at Riddlesdown Quarry; one was resting on a stem, with forceps raised in a threatening manner, while tiny nymphs ran around below her. They can be beaten from deciduous trees and occasionally pines and they will hide in almost any cavity; one was found in the cast skin of a dragonfly nymph on Bagmoor Common (RDH).

Richard Jones found a specimen of the macrolabic form, which has much longer forceps, in his garden at Nunhead in 1998. This form is reputedly uncommon inland but may be overlooked.

Adults are abundant in autumn; G.Fox-Wilson reported no less than 2148 individuals from beneath grease-bands on 22 apple-trees on 9 October 1930 at R.H.S. Wisley and Derek Coleman found hundreds in nest-boxes at Beddington in late summer. Males are particularly abundant at this time, whereas they are unknown in the spring. However Roger Hawkins kept a male alive in a refrigerator all winter, which suggests that they may survive the winter and mate in the spring before dying off, and Graham Collins beat two males from pines in mid-January.

Forficula lesnei Finot, 1887   PLATE 16   Lesne's Earwig

Nationally Scarce B

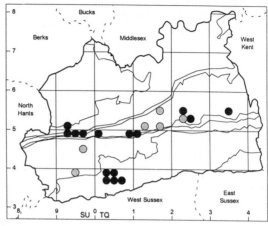

Lesne's Earwig is rather smaller and paler than the Common Earwig, being a rich red colour, and is distinguished also by having no visible wings. Once seen alive it is quite readily recognised, having a distinctive appearance. It is a nocturnal species and is usually only found by beating or sweeping old hedges, shrubs and vegetation on south-facing slopes of the Downs or other sheltered spots. Adults are found from late summer to November when the females, and possibly the males also, go into hibernation in hollow stems, such as those of hogweed. This earwig is widespread but local over southern England and on the Gower Peninsula, South Wales. It is certainly under-recorded and may in fact be reasonably common.

In Surrey Lesne's Earwig has almost certainly been overlooked and under-recorded, as elsewhere. It was first recognised in the British Isles in 1896 (having previously been misrecorded as *F. pubescens*) and in the following year it was found in Surrey by sweeping on the chalk downs near Reigate. In September and October 1898 Lucas found it commonly by beating bushes around Leatherhead. It was much commoner than the Common Earwig and he could find three or four specimens in his umbrella at the same time (Lucas, 1920). It was also recorded at this time from Box Hill, Ranmore and Witley. There are also two specimens collected in 1934 from Godalming in the O.H.Latter Collection at Charterhouse Museum.

There were then very few records until the last two years. Ian Menzies found it within the curled leaves of mullein in 1946 on Buckland Hills and in 1973 it was recorded from Abinger Hammer, Tadworth and Caterham (WRD). It was clearly still present on the North Downs but overlooked.

In September 1997 Roger Hawkins found it by beating hawthorn scrub at Blatchford Down above Abinger Hammer and then a week later, rather surprisingly, on weald clay in various sites between Cranleigh and Dunsfold; these he beat from hedges, scrub and thistles when searching for other insects. At the very end of the survey, in October 1998, it was found by beating, mainly from traveller's joy, in two places on the Hog's Back and at Pewley Down, Hackhurst Downs and Colley Hill (DWB, RDH, GAC, AJP) and also at Wanborough on London clay (RDH). At Hackhurst it was present in similar abundance to that found by Lucas a hundred years earlier at Leatherhead, i.e. three or four specimens in the beating tray at a time and much commoner than the Common Earwig.

Lesne's Earwig is probably common all along the scarp slope of the North Downs and possibly in suitable places off the chalk. It seems to prefer warm sheltered sites on the south-facing chalk slopes, rather than exposed sites. Off the chalk it probably prefers sheltered old hedgerows or rough vegetation and scrub on the south side of woods. It should be searched for by beating or sweeping in any likely spot, and all records should be notified.

APPENDIX 1 – Gazetteer of sites

Abinger Hammer	TQ0947	Bramley Bank, Selsdon	TQ3563
Addington Hills	TQ3564	Broadstreet Common, Guildford	SU9550
Addlestone	TQ0464	Brockham Quarry	TQ1951
Albury Heath	TQ0646	Brockwell Park, Brixton	TQ3174
Alfold Crossways	TQ0435	Brooklands	TQ0662
Arbrook Common	TQ1463	Buckland Hills	TQ2352
Arden Green, Lingfield	TQ3845	Burnt Common, Send	TQ0354
Ash Vale/Ranges	SU8953	Burwood Park, Weybridge	TQ1064
Ashtead Common	TQ1759	Byfleet	TQ0661
Auclaye Brickworks	TQ1738		
Backside Common, Guildford	SU9450	Caesar's Camp, Hale	SU8350
Bagmoor Common	SU9242	Camberley	SU8860
Banstead Heath	TQ2455	Camberwell Old Cemetery	TQ3474
Barnes Common	TQ2275	Carshalton	TQ2764
Battersea Park/Fields	TQ2876	Cartbridge, Send	TQ0156
Batt's Corner	SU8241	Caterham	TQ3355
Beddington	TQ2966	Charleshill, Tilford	SU8944
Berrylands	TQ1968	Charterhouse, Godalming	SU9645
Betchworth Park golf course	TQ1849	Cheam	TQ2463
Betchworth Quarry	TQ2051	Chertsey	TQ0466
Bisley	SU9559	Chiddingfold	SU9635
Black Down, Sussex	SU9129	Chipstead Valley	TQ2657
Blackheath, Chilworth	TQ0345	Chobham	SU9761
Blackwater Valley	SU8658	Chobham Common	SU9764
Blatchford Down	TQ0948	Churt	SU8538
Bletchingley	TQ3250	Clapham Common	TQ2874
Blindley Heath	TQ3645	Clasford Common, Guildford	SU9452
Boldermere, Wisley	TQ0758	Clockhouse Brickworks	TQ1738
Bookham Common	TQ1256	Coldharbour Common	TQ1444
Borough Farm, Milford	SU9241	Coleford Bridge gravel pit	SU8855
Botany Bay, Chiddingfold	SU9834	Colley Hill, Reigate	TQ2452
Bourne Woods, Farnham	SU8544	Colony Bog, West End	SU9259
Bowles Wood, Cranleigh	TQ0838	Compton	SU9548
Box Hill, Dorking	TQ1751		

99

Copthorne Common, Sussex	TQ3239	Fetcham, Leatherhead	TQ1554
Coulsdon	TQ3059	Folly Bog, Lightwater	SU9261
Cranleigh	TQ0638	Fox Corner, Worplesdon	SU9654
Croham Hurst, S. Croydon	TQ3363	Frensham Common	SU8540
		Friday Street, Dorking	TQ1245
Crooksbury Hill	SU8745	Frimley Green	SU8958
Croydon	TQ3265		
Croydon Airport (Roundshaw)	TQ3063	Gibbet Hill, Hindhead	SU8935
		Glover's Wood, Charlwood	TQ2240
Denmark Hill	TQ3275	Godalming	SU9743
		Gomshall	TQ0847
Devil's Punchbowl, Hindhead	SU8936	Greyspot Hill, Lightwater	SU9261
Dormans	TQ3941	Guildford	SU9949
Dormansland	TQ4042		
Dorking	TQ1649	Hackhurst Downs, Gomshall	TQ0948
Dulwich	TQ3372		
Dulwich Wood	TQ3371	Hagthorn Bog, West End	SU9260
Dunsfold	TQ0036	Hale	SU8448
		Ham Common	TQ1871
Earlswood	TQ2748	Hankley Common, Tilford	SU8841
Eashing	SU9443		
East Grinstead, Sussex	TQ3938	Hascombe	SU9939
Effingham	TQ1055	Haslemere	SU9033
Egham	TQ0071	Heaver's Meadow, Selhurst	TQ3367
Elm Park, Battersea	TQ2977		
Elstead	SU9044	Hedgecourt, Felbridge	TQ3540
Epsom Common	TQ1860	Hindhead Common	SU8936
Epsom Downs	TQ2158	Hog's Back	SU9248
Esher Common	TQ1262	Holmwood Common, Dorking	TQ1745
Esher golf course	TQ1465	Horley	TQ2942
Ewhurst Green	TQ0940	Horsell Common	TQ0060
		Hurtwood	TQ0843
Farncombe	SU9745	Hydon's Ball, Hambledon	SU9739
Farnham	SU8446		
Farthing Downs, Coulsdon	TQ3057	Itchingwood Common	TQ4150

Juniper Hall	TQ1752	Ockham Common	TQ0858
		Ockley	TQ1440
Kew Gardens	TQ1876	Ockley Common	SU9141
Kingston	TQ1870	Old Deer Park, Kew	TQ1775
Kingston Gate,		Old Woking	TQ0156
Richmond Park	TQ1970	Oxshott Heath	TQ1461
Knaphill	SU9658	Oxted	TQ3952
		Oxted Downs	TQ3854
Leatherhead	TQ1656		
Leith Hill, Dorking	TQ1343	Parkgate	TQ2043
Lightwater	SU9262	Peckham Rye	TQ3474
Limpsfield Common	TQ4052	Pewley Down, Guildford	TQ0148
Lingfield	TQ3843	Pickett's Hole, Ranmore	TQ1249
Lloyd Park, Croydon	TQ3364	Pudmore Pond, Thursley	SU9041
Lower Bourne, Farnham	SU8544	Purley Downs golf	
Lythe Hill, Haslemere	SU9232	course	TQ3261
		Putney Heath	TQ2373
Merrow Downs	TQ0249	Puttenham golf course	SU9447
Merstham	TQ2953	Pyrford	TQ0259
Merton	TQ2569		
Middle Bourne,		Quarry Hangers,	
Farnham	SU8445	Caterham	TQ3253
Milford	SU9441		
Millbridge	SU8442	Ranmore Common	TQ1450
Mitcham Common	TQ2868	Redhill	TQ2850
Moat Pond,		Reigate	TQ2550
Thursley Common	SU8941	Richmond Park	TQ1972
Morden Hall Park	TQ2668	Riddlesdown, Purley	TQ3260
Mytchett Lake	SU8954	Riddlesdown Quarry	TQ3359
		Ripley	TQ0556
New Cross Cutting	TQ3676	Roehampton	TQ2174
New Zealand golf course	TQ0360	Rotherhithe	TQ3579
Newdigate	TQ1942	Runnymede	TQ0072
Newlands Corner	TQ0449	Rushett Common,	
Nonsuch Park	TQ2363	Cranleigh	TQ0242
Normandy	SU9251	Ruskin Park	TQ3275
Nunhead Cemetery	TQ3575		

Location	Grid Ref	Location	Grid Ref
Sanderstead	TQ3361	Walton-on-Thames	TQ1066
Selborne, Hampshire	SU7433	Wanborough	SU9350
Selhurst	TQ3267	Wandsworth Common	TQ2773
Selsdon	TQ3562	Waterloo	TQ3179
Send	TQ0356	Waverley Abbey	SU8645
Shalford Meadows	SU9947	West End	SU9460
Shere	TQ0747	West End Common	SU9260
Shirley	TQ3564	West Ewell	TQ2163
Sidney Wood, Dunsfold	TQ0234	West Heath, Limpsfield Common	TQ4052
Smokejacks Brickworks	TQ1137	West Hill golf course	SU9656
Somerset Bridge, Elstead	SU9243	West Horsley	TQ0753
South Bank	TQ3080	Westcott Green	TQ1448
South Hawke, Oxted	TQ3753	Westfield Common, Woking	TQ0056
South Norwood	TQ3368	Weybridge	TQ0764
South Nutfield	TQ3049	White Downs, Gomshall	TQ1148
Southwark	TQ3280	Wimbledon Common	TQ2271
Sow Moor, Chobham	SU9861	Windsor Great Park	SU9770
St.George's Hill, Weybridge	TQ0762	Winterfold, Cranleigh	TQ0743
Stanhill Court, Charlwood	TQ2342	Wisley Common	TQ0658
Stoke Meadows, Guildford	TQ0051	Wisley R.H.S. Gardens	TQ0658
Strawberry Bottom, West End	SU9159	Wisley village	TQ0659
Stringers Common, Guildford	SU9952	Witley Common	SU9240
Stroude	TQ0069	Woking	TQ0159
Surbiton	TQ1867	Woldingham	TQ3756
Sydenham Hill	TQ3372	Wotton	TQ1346
		Wrecclesham	SU8245
Thursley Common	SU9040	Yateley Common, Hants	SU8359
Tilford	SU8743		
Tooting Bec Common	TQ2972		
Vauxhall	TQ3077		
Virginia Water	SU9969		

APPENDIX 2 – Organisations

Orthoptera Recording Scheme for Britain and Ireland
Publisher of the Newsletter: B.R.C., Institute of Terrestrial Ecology, Monks Wood, Abbots Ripton, Huntingdon PE17 2LS.
Editor and National Recorder: John Widgery, 21 Field View Road, Potters Bar, Herts EN6 2NA.

Surrey Wildlife Trust
School Lane, Pirbright, Woking, Surrey GU24 0JN.
County Orthoptera Recorder: David Baldock, Nightingales, Haslemere Road, Milford, Surrey GU8 5BN.

British Entomological and Natural History Society
Dinton Pastures Country Park, Davis Street, Hurst, Reading, Berks RG10 0TH.

Croydon Natural History and Scientific Society
96a Brighton Road, South Croydon, Surrey CR2 6AD.

DMAP enquiries to:
Dr.Alan Morton, Blackthorn Cottage, Chawridge Lane, Winkfield, Windsor, Berkshire SL4 4QR.

Fossil enquiries to:
Dr.E.A.Jarzembowski, Maidstone Museum and Art Gallery, St Faith's Street, Maidstone, Kent ME14 1LH.

APPENDIX 3 – References

Airy Shaw, H.K., 1945.
Platycleis grisea occidentalis Zeuner (Orth., Tettigoniidae) in Surrey. *Entomologist's Mon. Mag.* **81**:274.

Bellmann, H., 1985.
A Field Guide to the Grasshoppers and Crickets of Britain and Northern Europe. Collins.

Brown, V.K., 1973.
A key to the nymphal instars of the British species of *Ectobius* Stephens (Dictyoptera: Blattidae). *Entomologist* **106**:202-209.

Brown, V.K., 1983.
Grasshoppers. (Naturalists' Handbooks **2**.) Cambridge University Press. [Revised 1990 with new publisher – Slough: Richmond Publishing]

Burr, M.D., 1897.
British Orthoptera. The Economic and Educational Museum, Huddersfield.

Burr, M.D., 1902.
Orthoptera in the *Victoria County History of Surrey.*

Burr, M.D., 1936.
British Grasshoppers and Their Allies. Janson.

Burton, J.F., & Ragge, D.R., 1987.
A Sound Guide to the Grasshoppers and Allied Insects of Great Britain and Ireland. Tape cassette. Harley Books.

Collins, G.B., 1949.
Roeseliana roeselii (Hag). (Orth., Tettigoniidae) in Surrey. *Entomologist's Mon. Mag.* **85**:50.

Dandy, J.E., 1969.
Watsonian Vice-counties of Great Britain. Ray Society, London.

Edwards, M., Patmore, J.M., & Sheppard, D., 1996.
The Field Cricket – preventing extinction. *British Wildlife* **8**: 87-91.

Farrow, R.A., 1962.
Orthoptera of the Survey Area (Grasshoppers and Crickets). Croydon Natural History and Scientific Society. Regional Survey, Index No.51.

Haes, E.C.M., 1973.
　　The distribution of native Saltatoria in Sussex (1965-70). *Entomologist's Gaz.* **24**: 29-46.

Haes, E.C.M., 1976.
　　Orthoptera in Sussex. *Entomologist's Gaz.* **27**: 181-202.

Haes, E.C.M., & Harding, P.T., 1997.
　　Atlas of Grasshoppers, Crickets and Allied Insects in Britain and Ireland. The Stationery Office.

Hartley, J.C., & Robinson, D.J., 1976.
　　Acoustic behaviour in both sexes of the speckled bush-cricket *Leptophyes punctatissima*. *Physiol. Ent.* **1**: 21-25.

Lousley, J.E., 1976.
　　Flora of Surrey. David & Charles. Newton Abbot.

Lucas, W.J., 1920.
　　A Monograph of the British Orthoptera. London: Ray Society.

Lucas, W.J., 1921.
　　Notes on British Orthoptera, 1920. *Entomologist* **54**:95.

Mahon, A., 1988.
　　Grasshoppers and Bush-crickets of the British Isles. Shire Natural History series **25**.

Marshall, J.A., & Haes, E.C.M., 1988.
　　Grasshoppers and Allied Insects of Great Britain and Ireland. Harley Books. Colchester.

Martin, D.S., 1990.
　　The London Atalanta **17**:68.

Menzies, I.S., & Airy Shaw, H.K., 1947.
　　Roeseliana roeselii Hagenb. (Orth., Tettigoniidae), not *Platycleis occidentalis* Zeuner, in Surrey. *Entomologist's Mon. Mag.* **83**:151.

Newman, E., 1849.
　　The Letters of Rusticus on Natural History. London.

Paul, J., 1995.
　　Orthoptera in the London Archipelago. *Entomologist's Rec. J. Var.* **107**: 89-95.

Payne, R.M., 1958.
　　The distribution of grasshoppers and allied insects in the London area. *London Naturalist* **37** (1957): 102-115.

Pickard, B.C., 1954.
Grasshoppers and Crickets of Great Britain and the Channel Islands. Privately published.

Ragge, D.R., 1955.
Recent records of the Mole-cricket from Hampshire and Surrey. *Entomologist's Rec. J. Var.* **67**: 161.

Ragge, D.R., 1956.
Some notes on the Field Cricket *Gryllus campestris* L. (Orth: Gryllidae) *Entomologist* **89**: 300-301.

Ragge, D.R., 1965.
Grasshoppers, Crickets and Cockroaches of the British Isles. London. Warne.

Ragge, D.R., 1973.
The British Orthoptera: A Supplement. *Entomologist's Gaz.* **24**: 227-245.

Ragge, D.R., & Reynolds, W.J., 1998.
The Songs of the Grasshoppers and Crickets of Western Europe. Colchester. Harley Books.

Ragge, D.R., & Reynolds, W.J., 1998.
A Sound Guide to the Grasshoppers and Crickets of Western Europe. CD. Colchester. Harley Books.

Stephens, J.F., 1835.
Illustrations of British Entomology: Mandibulata **6**. London.

Sterry, P., 1990.
British Grasshoppers. *British Wildlife* **1**: 219-223.

Sterry, P., 1991.
British Bush-crickets. *British Wildlife* **2**: 233-237.

Stubbs, A.E., 1967.
The Wood Cricket, *Nemobius sylvestris* (Bosc) (Orthoptera: Gryllidae) in Surrey. *Entomologist* **100**: 284.

White, G., 1789.
The Natural History and Antiquities of Selborne. B.White & Son. London.

White, G., ed. Johnson, W., 1931.
Gilbert White's Journals. George Routledge & Sons. London.

APPENDIX 4 – Glossary

abdomen	the third, posterior part of the body.
adult	the final (mature) stage of life during which reproduction occurs.
beating	the striking of the lower branches of trees, over a footpath or tray, to dislodge insects.
cuticle	the skin of insects
frequency	the measurement of sound (in cycles per second or cps).
harp	a clear area of amplification on the forewings of crickets.
instar	a stage of growth in the life of an insect larva, terminated by a moult.
mirror	a specialised area of the forewing of male bush-crickets and crickets which acts as a resonator in stridulation.
nymph	the young stage(s) between egg (or worm-like larva) and adult.
ootheca	the case in which cockroach eggs are enclosed.
orthopteran	a member of the Orthoptera Saltatoria *sensu stricto;* the jumping Orthoptera with much-enlarged hind legs.
orthopteroid	one of the Orthoptera *sensu lato*; i.e. including the running Orthoptera and closely related orders without much-enlarged hind legs.
ovipositor	the egg-laying apparatus of a female.
palps	a pair of sensory structures under the mouth used in the testing and manipulation of food.
pronotum (adj. **pronotal**)	the dorsal surface of the first thoracic segment; in orthopteroid insects usually much enlarged
stigma	a white or pale-coloured spot on the forewings.
stridulation	the action of rubbing two parts of the body against each other to produce sound.

stridulatory pegs the small peg-like projections on the inner side of the hind femur of grasshoppers.

thorax the second major division of the insect body, bearing the legs and wings.

tibia the fourth segment and second long portion of the leg; the 'shin' of the the insect leg.

tympanum the thin membrane (the 'ear-drum') of the hearing organ.
(pl. **tympana**)

APPENDIX 5 – Plant names

Index of plants listed by common name

ash, *Fraxinus excelsior*

beech, *Fagus sylvatica*
bilberry, *Vaccinium myrtillus*
birch, *Betula pendula/pubescens*
bog asphodel, *Narthecium ossifragum*
bog myrtle, *Myrica gale*
box, *Buxus sempervirens*
bracken, *Pteridium aquilinum*
bramble, *Rubus fruticosus* agg.
buttercup, *Ranunculus* spp.

cherry, *Prunus* spp.
creeping thistle, *Cirsium arvense*
cross-leaved heath, *Erica tetralix*

false-acacia, *Robinia pseudoacacia*

gorse, *Ulex europaeus*

hawthorn, *Crataegus monogyna*
hazel, *Corylus avellana*
heather, *Calluna vulgaris*
hedge bindweed, *Calystegia sepium*
hogweed, *Heracleum sphondylium*
hornbeam, *Carpinus betulus*
horse chestnut, *Aesculus hippocastanum*

Lawson's cypress, *Chamaecyparis lawsoniana*
lime, *Tilia x vulgaris*

maple, *Acer* spp.
mullein, *Verbascum* spp.

nettle, *Urtica dioica*
Norway maple, *Acer platanoides*

oak, *Quercus robur*

plane, *Platanus x hispanica*
poplar, *Populus* spp.

ragwort, *Senecio* spp.
restharrow, *Ononis* spp.
rush, *Juncus* spp.

sallow, *Salix caprea/cinerea*
Scots pine, *Pinus sylvestris*
sycamore, *Acer pseudoplatanus*
sweet chestnut, *Castanea sativa*

thistle, *Cirsium/Carduus* spp.
traveller's joy, *Clematis vitalba*
tufted hair-grass, *Deschampsia cespitosa*
tussock-sedge, *Carex paniculata*

wavy hair-grass, *Deschampsia flexuosa*

INDEX

Figures in bold indicate plate numbers

Acheta domesticus 63,**8**
Apterygida media 95,**16**
Chorthippus
   albomarginatus 85,**12**
   brunneus 82,**11**
   parallelus 84,**12**
Conocephalus
   discolor 55,**5**
   dorsalis 58,**6**
Ectobius
   lapponicus 90,**15**
   pallidus 92,**15**
   panzeri 93,**15**
Forficula
   auricularia 96,**16**
   lesnei 97,**16**
Gomphocerippus rufus 87,**13**
Gryllotalpa gryllotalpa 70,**9**
Gryllus campestris 65,**8**
Labia minor 94,**16**
Leptophyes punctatissima 60,**7**
Meconema thalassinum 43,**3**
Metrioptera
   brachyptera 50,**4**
   roeselii 52,**4**
Myrmeleotettix maculatus 88,**14**
Nemobius sylvestris 68,**8**
Omocestus
   rufipes 80,**10**
   viridulus 81,**11**
Pholidoptera griseoaptera 48,**3**
Stenobothrus lineatus 78,**10**
Stethophyma grossum 77,**10**
Tetrix
   subulata 74,**9**
   undulata 75,**9**
Tettigonia viridissima 46,**3**

Bog Bush-cricket 50,**4**
Common Earwig 96,**16**
Common Green Grasshopper .. 81,**11**
Common Groundhopper 75,**9**
Dark Bush-cricket 48,**3**
Dusky Cockroach 90,**15**
Field Cricket 65,**8**
Field Grasshopper 82,**11**
Great Green Bush-cricket 46,**3**
House Cricket 63,**8**
Large Marsh Grasshopper 77,**10**
Lesne's Earwig 97,**16**
Lesser Cockroach 93,**15**
Lesser Earwig 94,**16**
Lesser Marsh Grasshopper 85,**12**
Long-winged Cone-head 55,**5**
Meadow Grasshopper 84,**12**
Mole-cricket 70,**9**
Mottled Grasshopper 88,**14**
Oak Bush-cricket 43,**3**
Roesel's Bush-cricket 52,**4**
Rufous Grasshopper 87,**13**
Short-winged Cone-head 58,**6**
Short-winged Earwig 95,**16**
Slender Groundhopper 74,**9**
Speckled Bush-cricket 60,**7**
Stripe-winged Grasshopper 78,**10**
Tawny Cockroach 92,**15**
Wood Cricket 68,**8**
Woodland Grasshopper 80,**10**